ANIMAL WONDERS
of the WORLD

ANIMAL WONDERS
of the WORLD

David Black

Orbis Publishing · London

First published in Great Britain
by Orbis Publishing Limited, London 1981

Printed in Czechoslovakia

ISBN: 0-85613-391-4
50151

Title page: Despite its ferocious
appearance, the mountain gorilla
(*Gorilla gorilla berengei*) is a gentle animal
existing solely on a vegetable diet. Its
great strength is seldom put into effect
and is used only as a threat against other
males and to defend its troop against the
occasional marauding leopard.

Acknowledgements: Ardea 20t, 21b,
53, 54, 88, 94, 108tr, 109t, 109b, 119,
147b, 151b, 167, 197; C. Bevilacqua 22,
23, 165; Borgioli and Cappelli 59, 60t,
60c, 60b, 177t, 177b; M. Cedri/Petron
174t; Bruce Coleman 2, 10, 11, 12t, 12b,
15t, 16-17, 19, 21t, 29, 36-37, 42, 44-45,
46b, 51, 67, 69t, 70, 74-75, 77, 80-81, 87,
89, 93, 99, 110, 117, 120-121, 122t,
128-129, 135, 141r, 142, 151t, 152, 153,
160-161, 168-169, 171, 173, 180-181,
183, 184, 185, 192b, 194; S. Y. Craig 83;
F. Erize 130, 131; D. Faulkner 143;
M. Fogden 71b, 137b, 140; Eric Hosking
7, 86t; ICP 145, 146; Achivio IGDA
84-85, 91tl, 91tr, 141l; Archivio
IGDA/A. Margiocco 134b; Jacana 8-9,
13, 14, 15b, 26, 27t, 27b, 30, 31, 32, 33,
49, 64-65, 66, 68, 69t, 69b, 76, 86c, 86b,
97, 100, 101b, 104-105, 111, 112, 113,
114, 115, 116, 122b, 123, 163, 172, 175,
187, 189, 191, 195, 199, 200; Frank Lane
156l, 157; G. Mazza 91tl; C. Mylne 92;
NHPA 90l, 90r, 91b; Natural Science
Photos 43; John Norris Wood 133;
Oceanic Press/Petron 147t; Popperfoto
134t; Seaphot 54-55; J. Six 58t; Peter
Ward 40-41, 73, 132, 136, 137t, 138-139;
World Wildlife Fund 39; I. Wyllie 107;

Contents

INTRODUCTION

The diversity of animal life is truly astounding when one considers its range of size and form, from the microscopic protozoa through the vast assemblage of invertebrate animals – the soft-bodied sponges, jellyfish and molluscs – to the insects, the most numerous of all species. Birds and mammals, the most familiar animals, make up only a tiny fraction of the total number of animal species – of which over 1,071,000 have so far been identified.

Animals are found throughout the world's varied habitats from the polar seas and windswept mountain tops to the tropics, where the greatest diversity of life is found. In the hothouse atmosphere of the tropical forests plants and animals have evolved into a staggering variety of forms. Here insects resemble flowers, harmless snakes mimic deadly ones and tiny frogs that hop about the forest floor contain enough poison within their bodies to kill 20,000 men. Vying with the tropical forest in terms of abundance of life are the coral reefs whose massive structures have been built up over thousands of years. Man is just beginning to unravel the complicated relationships which exist between the multitude of marine life forms that are found in this astonishing habitat.

While Western man becomes more alienated from the natural world and views the wonders of animals as a sophisticated voyeur, to the peoples of the Third World animals are a constant and real threat. Man's minute enemy the mosquito remains unconquered, causing sickness and death on a scale difficult for us to comprehend. Snake-bite kills thousands of people each year in Asia; in Mexico scorpions constitute a serious hazard; while in Africa, apart from the insects that transmit disease, people's livelihood is periodically decimated as their crops are stripped bare by dense swarms of locusts and also by marauding quelea, which during their search for food, darken the skies with their vast flocks. On the other hand, man himself through his population expansion, hunting and sheer greed has been responsible for the extermination of many species. The dodo, passenger pigeon and Tasmanian wolf are more familiar examples of man's destruction. Other animals, like the Arabian oryx, have been rescued at the last minute and now consist solely of captive-bred specimens.

This book is structured to give as wide a range as possible of animal diversity. Although one is unlikely to witness the breaching of a blue whale, the largest of all animals, or a rare snow leopard at its kill, many of the animals featured can be seen close to home or within a short walk into the countryside. For example, on a mild spring morning after rain, garden snails can be found going through the incredible courtship ritual which is described in detail in this book. The pugnacious territorial battles of the robin, one of the shortest-lived of birds, can be easily seen in many gardens, while one may have to venture further afield to monitor the regular arrival each spring of the cuckoo. If one is patient, it is not a difficult task to track down the female bird and find the nests of those songbirds she has so effectively parasitized.

The formidable armaments, speed, slowness or strength, poisons, gaudy colours or cunning camouflage exhibited by the animals featured in this book are just some of the many animal wonders that have evolved not for our benefit as human spectators of wildlife, but as aids to the battle for animal survival.

THE GREAT
—AND—
THE SMALL

Gathering of the giants – a herd of
African elephants *(Loxodonta africana)* drink at
a water hole. The African elephant is the largest
land animal. An average bull stands
over 3 metres (10 ft) at the shoulder
and weighs 2.5 tonnes.

The Gorilla

Gentle giant of the forest

Captive gorillas have been known to burst footballs under their arms, send stationary cars crashing against walls and lift enormously heavy weights with ease, yet in the wild these giant apes lead a peaceful life existing solely on a vegetable diet of stems, shoots and roots. Of the three sub-species, the eastern lowland gorilla (*Gorilla gorilla graueri*) is the largest; an adult male weighs 160 kilograms (352 pounds), stands 1.7 metres (5 feet 7 inches) tall and has a massive chest girth of 1.5 metres (59 inches). Males grow much bigger than females, but their greater bulk and muscular power is only used as intimidation against males from other groups or against another large mammal such as man or a predatory leopard.

Attacks on man have occurred, but in all cases the gorilla in question was either suddenly surprised or deliberately provoked into attack. Some ten years after the discovery of the gorilla in 1847 by an American missionary, the eminent British anatomist Richard Owen wrote of it:

> Negroes when stealing through the shades of the tropical forest become sometimes aware of the proximity of one of these frightfully formidable apes by the sudden disappearance of one of their companions, who is hoisted up into a tree, uttering perhaps a short croaking cry. In a few minutes he falls to the ground a strangled corpse.

No doubt Owen had been misled about the gorilla's true nature by the many tall stories brought back by missionaries and hunters from the African forests. It seems that man likes to harbour a fearsome, larger-than-life animal in his imagination; a hundred years ago a mysterious man-killing ape fitted the bill, while today we are bombarded with more sophisticated images of terrifying aliens.

Diet and Social Organization

Today the gorilla is an object of curiosity and concern as its forest habitat is rapidly eroded. The rarest and best-studied sub-species is the mountain gorilla (*G.g. beringei*) which is confined to the forested slopes of mountains and volcanoes in the central African countries of Rwanda, Uganda and Zaire. In this lush and humid landscape the gorilla finds numerous varieties of edible plants. Even so, it is highly selective. When eating a plant of wild celery it will delicately remove the leaves and outer part of the stem before extracting the pith. When whole plants are uprooted for their tubers and roots, a gorilla will carefully shake them to remove all the earth. Stinging plants such as nettles, and plants with prickles and spines are all eaten with apparently no ill effects although it takes young animals some time to master the technique used for a particular type of vine which is armed with large hooks. Adults deftly roll these up into a ball before eating. A gorilla's succulent diet provides it with all the water it requires. An adult male is calculated to eat 30 kilograms (66 pounds) of food every day which contains over 25 litres (44 pints) of water.

Mountain gorillas live in groups of from five to thirty individuals led by a single large

Right: An adult gorilla may weigh over 160 kilograms and have a chest measurement of over 1.5 metres.

Left: An immature mountain gorilla (*Gorilla gorilla beringei*). At seven months old a gorilla can climb up and down trees with great agility although adults spend most of their time on the ground.

adult male known as a silverback. The name derives from a prominent patch of grey hairs which develops at maturity as a saddle-shaped patch on the back and spreads with age to the buttocks and flanks. Whatever decision the silverback takes the rest follow and that includes other sub-adult males as well as additional but subordinate silverbacks. When a group moves through the forest the dominant silverback takes the lead and a second silver-

back will take up the rear. Fighting between males is rare but a silverback will sometimes show his dominance by exhibiting his famous chest-beating display which is often mimicked by other members of the group. Actually chest beating is just one of a sequence of nine actions which make up a gorilla's intimidation display. The sequence involves hooting, symbolic feeding, standing up on the hind legs, throwing vegetation, chest beating, leg kicking, running, tearing vegetation and thumping the ground. This all lasts for less than a minute but is in sharp contrast with the gorilla's normally slow and unhurried movements.

Gorillas are creatures of routine, waking around dawn and feeding continuously for a few hours, then around noon sleeping until mid-afternoon when they continue foraging for food which may at this time take them some distance into the forest. An hour or so before dusk they settle for the night, each gorilla building itself a rudimentary nest. The silverback initiates the nest building by twisting and tearing off great bunches of vegetation. Females with young will often build a platform of leaves and branches low down in the fork of a tree. The dominant male may doze off with other gorillas lying up to him in a bunch.

Like other apes the female gorilla is sexually receptive throughout the year. Apparently a female will mate with any male of the group, and in the presence of the others. It is she who normally takes the initiative by crouching down and backing up towards him displaying her genitalia, all the time looking back towards him. During copulation both sexes have been heard to make gentle cooing sounds and afterwards may lie together quietly for half an hour or so. The female gives birth to one or occasionally two young every three or four years. The babies are carried around for three months and after this take their first steps. By seven months they are extremely agile, can climb trees, swing from branches and play with other young members of the group. Juveniles and adolescent gorillas spend a lot of time playing around; their favourite games include sliding down greasy slopes or walking around in a comical fashion with clumps of leaves on their heads.

Man and the Mountain Gorilla

Whilst several zoologists have lived amongst mountain gorillas giving us endearing pictures of these apes and recording interesting aspects of their behaviour, their future survival is doubtful unless a massive security operation is mounted and the local populations take up the challenge to save this unique animal. Unfortunately the distribution of the mountain gorilla spans three countries whose peoples have shown little interest in the animal's protection. Gorillas are hunted for their meat, for the skulls and hands which are made into grotesque tourist souvenirs, and in parts of their range gorilla finger bones are coveted as a supposedly powerful aphrodisiac. The zoo trade has not helped either as a juvenile gorilla is worth at least 6000 dollars to animal traders and this type of money is a big incentive in an area where most of the population live on subsistence level. At the same time the gorilla's habitat is dwindling as the forests are chopped down, crops grown and cattle browsed.

The most recent move made towards the gorilla's protection is the Fauna Preservation's Mountain Gorilla Project whose target is to save the 300 surviving mountain gorillas in Rwanda's Parc de Volcans centred around the Virunga craters. This park is a great improvement on cramped zoo conditions.

Right: Gorillas are great wanderers, rarely spending more than a day in one place. They are not known to drink, obtaining all their water from their succulent plant diet.

Below: Gorillas usually walk about on all fours, their body weight being supported mainly by the knuckles of the hands.

Bottom: A gorilla's intimidation display starts with him stuffing vegetation into his mouth and finishes with him banging the ground with the palms of his hands.

12

The Etruscan Shrew

Mammalian midget

All shrews are small, but beside *Suncus etruscus*, the Etruscan, or Savi's pygmy shrew, all others are giants. The tiny common shrew (*Sorex araneus*) is about twice its size and the familiar water shrew (*Neomys fodiens*) is even larger. An Etruscan shrew may measure as little as 36 millimetres ($1\frac{2}{5}$ inches) from the tip of its nose to the root of its tail and its tail may add another 24 millimetres (1 inch). It weighs up to 2.5 grams (0.09 ounces). It is the smallest known *terrestrial* mammal, the smallest mammal of all being the rare bumble-bee bat of Thailand, which has a wingspan of 1.6 centimetres ($\frac{3}{5}$ inch) and weighs up to 2 grams (0.07 ounces). Should a mammal smaller than these two ever be discovered, it would only be very slightly smaller. This is because it would not be possible for a mammal much smaller than the Etruscan shrew to survive at all. Oddly enough, it would starve.

A mammal must maintain a constant temperature inside its body and for much of the time this temperature is different from the temperature of the surrounding air. The animal must keep itself warm or cool. To do this it must use energy, which it obtains from the food it eats, and the amount of food used in this way is considerable. By basking in the sun to warm itself and seeking shade or water to cool itself, a reptile can manage on about one-tenth the amount of food needed by a mammal of the same size.

Depending on whether they are warmer or cooler than their surroundings animals gain or lose heat through their skins. The amount of heat that is exchanged in this way depends on the amount of skin – or surface area – they have compared to the volume of their whole bodies. Clearly an animal with a large body surface and small volume will have to work harder to maintain a constant internal temperature than an animal with a large volume and small surface area. The bigger an animal is, the larger will be its volume and the smaller its surface area in proportion, and so the less food it will need. This explains how whales can live comfortably in water that is almost freezing, why cows are more concerned with keeping cool than with keeping warm, and why shrews have large appetites.

In order to maintain its internal temperature a shrew must eat almost constantly. It must feed by night and by day – although by day it tries to remain hidden – and it can sleep only in short naps, from which hunger awakens it. An adult eats about two-thirds of its own body weight in food each day, and it can starve to death if it goes without food for more than about three hours. Its death in this case is due not to the wasting away of its body, but to its loss of control of its temperature so that it dies from heat or cold. Shrews are obviously very successful at staying alive, but a mammal that was appreciably smaller than the Etruscan shrew simply could not eat quickly enough to prevent itself from dying.

To some extent the shrew helps itself by living only where the climate is mild. The Etruscan shrew is found only in southern Europe, in the countries that border the Mediterranean, and in the Asian subtropics. Even there, though, nights can be cold and winters harsh.

Lifestyle and Physiology

Shrews are insectivores. The term describes their structure, especially their teeth, rather than their actual diet, for a shrew will eat any small invertebrate it finds, as well as some seeds and other vegetable matter. It moves rapidly,

Right: The tiniest of terrestrial mammals is the Etruscan shrew (*Suncus etruscus*) of southern Europe and Africa. It measures less than 4 centimetres ($1\frac{1}{2}$ inches).

in short bursts punctuated by pauses. Not much is known about the habits of the Etruscan shrew, but probably it swims well – most shrews do – and climbs well, but only if it has to. It lives a solitary life, except when females are raising young, nesting among the roots of trees – especially the small cork oak (*Quercus subur*) that grows in southern Europe – and hunting among the shrubbery. It does not welcome visits from other shrews.

Indeed, shrews are proverbially bad-tempered, and encounters between individuals are noisy, quarrelsome affairs. They do not lead to injury, however, the quarrels consisting only in the trading of insults. Eventually the loser retires and is not pursued. Where even the briefest interruption in the food supply can spell disaster there are advantages in keeping the population dispersed, and for the busy shrew there is no time to waste in fighting. Shrew mothers, on the other hand, are attentive and protective toward their young.

All shrews produce a weak venom. It is strong enough only to subdue an animal the size of an earthworm, but it serves as a defence for all that, because the carnivore that eats a shrew usually vomits at once. Shrews are eaten, but only by predators that are desperately hungry. More often they are killed, probably because they have been mistaken for mice, and left uneaten. This production of venom is a feature we associate more usually with reptiles and it is one of the reasons for supposing that the shrews evolved a very long time ago – their history has been traced back about eighty million years – and are not too different from the earliest mammals. It is believed that several groups of mammals, including the bats and ourselves, are descended from shrew-like ancestors.

The Etruscan shrew can be recognized by its very small size, but also by the very long hairs that grow among the short hairs of its fur, especially on its tail. Its teeth are white, its ears large and erect, and there is a reddish tinge to the grey-brown fur of its back.

The Albatross
Long-distance sailor on the wind

Main picture: The royal albatross (*Diomedea epomophora*) has a wingspan of up to 3 metres (10 feet). It breeds on islands in the Southern Ocean near New Zealand.

The wandering albatross (*Diomedea exulans*) is one of the most beautifully proportioned of all flying machines. Its wingspan of over 3 metres (10 feet) is larger than that of any other bird and enables it to glide for months on end with a wonderful, characteristic sweeping motion up and over the waves far out at sea. This giant of the albatross tribe lives in the southern oceans between the 40th and 60th parallel, an area which encompasses that notoriously turbulent belt known as the 'Roaring Forties' where the continuous gusts of winds and high swell seas provide the albatross with the wind energy it needs to maintain its timeless, gliding flight.

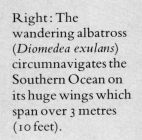

Right: The wandering albatross (*Diomedea exulans*) circumnavigates the Southern Ocean on its huge wings which span over 3 metres (10 feet).

Far right: The yellow-nosed albatross (*Diomedea chlorothyncus*) frequents the South Atlantic and Indian Ocean.

Right: the black-footed albatross (*Diomedea nigripes*) nests on islands in the North Pacific.

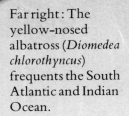

By travelling across the direction of the wind the albatross takes advantage of the different wind speed at different heights above the surface. As it plummets down towards the water its wings form a shallow 'W', which reduces drag; just before it hits the surface it suddenly veers along a wave trough gaining sufficient lift from the air currents formed at the wave crest to send it buoyantly aloft again. Using this thrust together with the energy gained from its dive it rises to about 15 metres (50 feet) before swooping down again. All the time the albatross scans the surface waters for food such as squid and fish which are snapped up by the hooked serrated bill before continuing its rhythmic flight pattern. With such mastery of the air with the minimum expenditure of energy, an albatross can fly thousands of kilometres without visiting land, although it may intermittently rest on the sea for short periods. Its ability to cover enormous distances was verified by the recovery of one tagged individual bird which was found over 4000 kilometres (about 2500 miles) away from its breeding ground where it was known to be feeding its chick. The related Laysan albatross (*D. immutabilis*) was timed as flying 5150 kilometres (3200 miles) across the North Pacific in just ten days.

Breeding and Survival

All the breeding sites are isolated islands including Inaccessible and Gough Islands in the South Atlantic and Marion and Prince Edward Islands in the southern Indian Ocean. During early summer (November in the southern hemisphere) one by one the albatrosses head for their breeding grounds – often making awkward crash-landings as they hit the short windswept turf. The first activities involve building the nest which is a volcano-shaped mound of turf about 30 centimetres (1 foot) in height. Once this is completed a pair perform an elaborate courtship display. This involves them facing each other with their long wings outstretched, mutual bowing and an ecstatic cacophony of bill clattering. A single large egg weighing approximately 450 grams (1 pound) – equivalent to six large hen eggs – is laid towards the end of December. The egg is incubated for seventy days by both parents. Once hatched the chick is fed intermittently for up to ten months. It gains weight at the fast rate of 100 grams (3½ ounces) a day and rapidly changes its white down for brownish juvenile plumage, and then full adult garb at about a year old. As the bird grows it is left for longer periods and must defend itself from frequent attacks by marauding skuas. Once it leaves the nest a young bird may not return to its home island to breed for several years.

Of the other twelve species of albatross the royal albatross (*D. epomophora*) is only slightly smaller and is virtually indistinguishable from the wandering albatross in flight, both being pure white with black wing-tips. In the North Pacific there are three species ranging between Japan and the western seaboard of North America. The Laysan and black-footed albatross both breed on the sub-tropical islands around Hawaii while the third species, the short-tailed albatross, breeds on the Japanese archipelago. All three have the curious habit of laying their eggs during late autumn and winter which may be to relieve the chick of the hot summer sun but is probably more likely to be tied in to the availability of food.

Below: An albatross travelling across the direction of the wind glides for days on end without flapping its wings by using the different wind speeds above the water surface.

The Hummingbird
Tiny rainbow of the jungle

A hummingbird is a living paradox. It is the smallest of birds, yet its performance in many aspects is more remarkable than that of birds which dwarf it into insignificance. Not that a hummingbird would ever claim less than one's full attention. Its acrobatic flight and the startling beauty of its plumage command no less; but, beyond the obvious, its anatomy and capabilities outstrip those of many birds that are widely considered champions in their class. Any account of this tiny creature rings with superlatives.

Doubtless it was the brilliance and beauty of the hummingbirds' plumage and their diminutive size that first captured the attention of European biologists in the last century. Although American ornithologists and a fortunate few Europeans who had travelled in the Caribbean and Central America had long known of these delightful creatures, the hummingbirds did not make their debut before entranced zoo visitors in Europe until early in the present century. Then, briefly, the first hummingbird seen alive in Europe stared at a spellbound audience from its cage in the Zoological Gardens in London. Its appearance was a short one, for the captive was little understood, and soon died. It lived long enough to fire the ambitions of taxidermists and milliners with new ideas for decorative set pieces of dead hummingbirds arranged among foliage within glass domes for middle-class mantelpieces, and hats ablaze with gem-like hummingbird feathers. The trade in the birds was a successful one. Taxidermists were pleased to find that the birds had tough skins, and the milliners were relieved to find that the feathers retained their astonishing brilliance when detached from the birds.

In their native forests, the hummingbirds withstood the assault on their numbers until the fashion for their corpses declined. Their smallness, agility and toughness make them difficult to wipe out entirely. Their size becomes an advantage in the fight for survival when it is matched by a skill in flight that is second to none and by an aggressive determination that seems more fitting in terriers than in miniature birds.

There are more than 320 species of hummingbirds in the world, and their size varies from minute to very small. The giant hummingbird (*Patagona gigas*) is about the same size as a swallow, about 21 centimetres (8½ inches) long, but more than half of this length consists of bill and tail feathers. It weighs little more than half the weight of a canary. The smallest of hummingbirds is the female Cuban hummingbird (*Mellisuga helenae*), which measures no more than 5.5 centimetres (2⅕ inches) from the tip of its bill to its tail. Its bill is 1.5 centimetres (⅗ inch) long, and its tail is just over 2.5 centimetres (1 inch), leaving a body of 1.5 centimetres (⅗ inch) into which a miracle of organs and muscles is packed – and all this without the benefit of transistors.

Anatomy and Plumage

The hummingbird's flight muscles account for about a quarter of its weight: a larger proportion than in any other bird. They are attached to a well-developed breastbone with a deep keel to give them good anchorage. The wings themselves are unusual. The humerus, or upper arm bone, is short, and the wings are supported mainly by the wrist and hand bones. The bird can swivel the wing bones at its shoulder-girdle so that the wings turn through almost a semicircle. It is this unique gift that enables the hummingbird to perform its most amazing feats of flight.

Most birds perform a wasted action in flight when, after the downbeat of their wings has provided power, the upstroke makes no further contribution. Hummingbirds at the end of the downstroke tilt the trailing edges of their wings further down so that, as they recover their original position, they are pushing air back again, providing a forward thrust. The ability to alter their wing positions so drastically produces an agility in flight that is matchless. No other birds can hover so steadily – the hawks look clumsy and flurried when compared with a hovering hummingbird – or change direction so crisply. The hummingbird makes right-angled turns and stops dead in mid-flight in a way that would rip the wings

from a modern fighter plane if its pilot were to attempt similar manoeuvres.

To watch these small birds dancing their way between the stems and branches of the forest, pausing and tacking from flower to flower, suddenly rising vertically or swooping in a powered dive – a hummingbird is not content to simply close its wings and fall, it works at its descent with blurred wings – is a hypnotic experience. Their wings beat a barely credible sixty to eighty times a second to make the characteristic hum that explains the bird's name. In fast flight, the wings become almost invisible, and when in aggressive movement – and they fight often – the hum becomes a furious buzz.

As hummingbirds change direction in flight, their plumage may appear at one moment a

Above: The glittering-bellied emerald hummingbird (*Chlorostilbon auroventris*) is one of eleven species characterized by their overall metallic plumage.

Below : A hovering
hummingbird burns
energy at a rate ten
times as fast as a
running man.

Bottom : A
hummingbird's
variable flight
patterns : A and B
forward propulsion ;
C hovering, where
the wings describe a
figure-of-eight. Up
to ninety wing-beats
per second have been
recorded for the
South American
species *Heliactin
cornuta*.

dull dark brown, and the next illuminated by a jewelled radiance. This mysterious change is due to the structure of the feathers, the finest branches of which consist of minute hooks called barbules which link to hold the vanes of the feathers together in an orderly way. A hummingbird's feathers, like those of some other birds with iridescent plumage, have a coating of lamellae, which are myriad microscopic plates of a horny substance known as keratin. The lamellae of some hummingbirds are only 0.00025 millimetres thick, but they are elliptical in cross-section and filled with pockets of air. As the sunlight strikes them, they refract light and reflect gold, blue, green or violet light in rapidly altering proportion as the bird changes its angle to the observer.

The tight-packed closeness of a hummingbird's plumage makes some parts of its body

seem to be of a continuous sheet of metal stained with gorgeous rainbow colours, but the bird in fact has denser plumage than any other bird there is. A hummingbird bathing is a wonderful sight, for as the bird flutters its wings in the dew pool of a tropical leaf, its colours may change from predominantly blue to mostly red, always with scatterings of rainbow colours among them. The water on the lamellae has the effect of increasing the thickness of the tiny plates – the water, in a sense, adds an extra 'lens' to them – and changes the wave-length of the refraction, so altering the bird's predominant colour.

Feeding

The intense activity that characterizes the hummingbird's life is fuelled by a specialized diet. The bird's metabolic rate is extremely high so its food must be rapidly digested and of concentrated nourishment, rich in energy sources. A hummingbird's principal food is nectar gathered from flowers. The nectar is an excellent source of glucose and fructose, which is quickly converted to fuel for flight and the maintenance of the body's functions. The proteins, fats, minerals and vitamins needed to keep the bird healthy come from the insects that it catches in flight. In an average thirteen-hour active day a bird has to consume a great deal of this high-energy food. On most days it visits about 2000 flowers to find the nectar it requires. At each flower, it hovers so that it can insert its bill, perfectly adapted for the purpose, between the petals, thrust out its long, channelled tongue and draw up the nectar.

Many of the flowers hummingbirds feed from have evolved without the 'landing platform' petals found in most flowers that are pollinated by insects. Hummingbirds pollinate these flowers as they feed without having to land on the bloom itself. Some flowers have their nectar hidden deep inside their corolla, and the hummingbirds who feed from them regularly have developed long, curved bills ideal for the purpose. Other hummingbirds, impatient with the problem of winning their nectar in the usual way, drive their bills into a protecting petal and rotate rapidly, like an avian drill, make a neat hole, and suck the nectar through it. Generally, the birds do little damage to the blossoms, and serve the flowers well by spreading their pollen.

The shape of the bill is important to the hummingbird's style of feeding, but equally

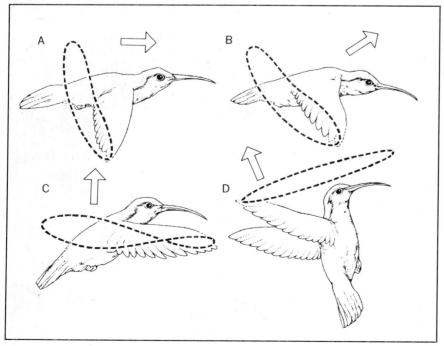

important is the highly specialized tongue. This, in many of the birds, may be extended far beyond the end of the already long bill. The tongue is attached by muscles that fold round the back of the skull, and when the bird thrusts its tongue forward, the muscular movement can be seen through the feathers behind the bird's head. The tongue itself is white, and grooves run along its length to the tip which divides into a vee shape, a little like a snake's. The inner edges of the vee are fringed, and the bird seems to use this area to draw nectar into the grooves along the blade of its tongue.

Nectar and a few insects might seem a small supply of food for this energetic bird, but multiplying the rate of energy exchange and the food intake to find a human equivalent suggests that a man would have to eat around 115 kilograms (250 pounds) of meat a day to match the hummingbird. Most of the birds eat almost continuously each day, but migratory species, such as the ruby-throated humming-bird (*Archilochus colubris*), eat furiously before their long trek – in this instance, from Alaska to Mexico – to build up a reserve of fat that will supply them with the energy they need for the journey.

Nest and Rest

The nests of hummingbirds are among the most beautiful of birds' homes. Their small-ness, the neatness of their construction, and the improbability of their position in some cases makes them especially interesting. The nest usually fits the female bird and her two eggs exactly. It is finely woven of delicate strands of fibres and spiders' silk. The female, who makes the nest on her own, lines it with moss. She often chooses a site where the approach is difficult. Observers have frequently described nests in such awkward places that the bird has to hover outside the nest's entrance, turn round, and reverse in.

The chicks stay in this beautifully wrought nest for only twenty-three days before whirr-ing away on their maiden flight. Although these minute creatures are born blind and helpless, they quickly grow their marvellous armour of feathers, missing out that downy stage common to most young birds. It is the female hummingbird who takes full responsi-bility for nest building and the raising of the young birds. The mate, after a breathtaking courtship display of aerobatics, has very little further contact with his family.

In the dying light of evening, the birds roost. Their feet firmly grasp a perch and they fluff out their feathers. As they rest, their metabolism slows down so that they assume a state that is rather like hibernation. Their heartbeats slow down and they lose heat. There they cling until roused by the warm sun of the following morning, when, whirring their wings, they rise vertically from their roosts like miniature, heraldic helicopters.

Below: A long-tailed hermit hummingbird (*Phaethornis superciliosus*).

Bottom: A 9-centimetre ($3\frac{1}{2}$ inch) copper-rumped (*Amazilia tobaci*).

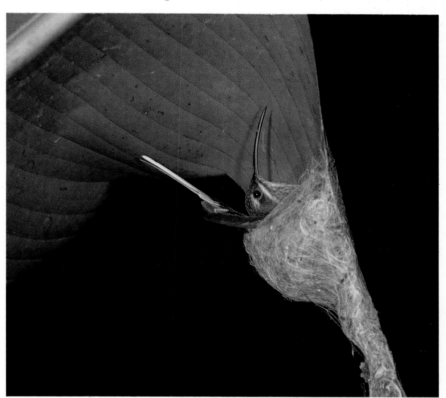

Giant Beetles

Winged heavyweights of the insect world

The largest of the 275,000 species of beetle include members of two sub-families – the Cetioniinae and the Dynastinae. Among the Cetioniinae, the goliath beetles (*Goliathus sp.*) of Africa are the heaviest, weighing up to 100 grams ($3\frac{1}{2}$ ounces) and measuring up to 12 centimetres ($4\frac{3}{4}$ inches). These insects are also strikingly marked chocolate brown or black streaked with white on the foreparts, and the whole body is clothed with a beautiful velvet-like 'fur' which, unfortunately, comes off with handling. Goliath beetles are found in the tropical equatorial forests of Africa, where they feed on pollen, fruit and sap while the massive cylindrical larva feeds on the decaying wood of the large forest trees. They are day-flying insects and are commonly seen grasping a branch with all six legs or else hanging down gibbon-style acrobatically by one leg. When threatened, the goliath beetle has an unusual defence mechanism – it lowers the front part of its thorax, revealing a wide, sharp-edged joint between the prothorax and the wing covers. This can be snapped shut powerfully and tightly on any animal such as a small monkey that is tempted to snatch it as a meal. The first time these wonderful beetles were displayed in Britain was in 1898 at the London Zoo where evidently they thrived on a diet of melons.

In the same sub-family, one finds the common rose chafer (*Cetonia aurata*), a beautiful metallic green-red beetle that can be seen during the summer eating rose-petals, and later in the season on fallen fruit.

Battle of the beetles

The other giants belong to the sub-family Dynastinae and include *Dynastes hercules* which grow up to 20 centimetres (8 inches), most of which comprises the enormous horns which have given them the name of rhinoceros beetles. This species from South America is nocturnal and feeds on vegetable matter, using its horns mainly in combat with other rival males which takes place at the onset of the rainy season. The antagonists face each other and make jerky movements trying to joust with their horns to raise the other off the ground – if one succeeds, the loser is sent crashing to the ground. Evidently, small males, because they are more nimble, often succeed against more bulky opponents. The victor invariably marches stiff-legged round his fallen foe.

The fighting between males of the related elephant beetle, *Megasoma*, was observed by the American zoologist, William Beebe:

> The opponents meet head on and either wait warily for the other to attack or one may rush headlong and begin the encounter. Usually both wait and spar at a distance. The object first noticeable is an attempt with one or both tarsi [feet] and claws to trip and unbalance the opponent. . . There are quick forward lunges and reachings out with one or both legs. . . This may or may not succeed but one will eventually force the fighting and there is straight pushing and butting for a considerable period, exactly like two antlered deer. Now and then, an effort will be noticed to lower the head and get the cephalic horn beneath the other insect. . . Periods of rest or waiting may intersperse the encounter and twice I have seen one beetle turn and rush after the female. In both cases the other was after

Right: The longest beetle is *Dynastes hercules* which grows up to 20 centimetres (8 inches). Over half of this comprises the large horns which are used by the males in combat.

Right: The beautifully marked *Goliathus regius* of equatorial Africa is the heaviest known beetle, weighing up to 100 grams (3½ ounces).

him at full speed and the battle began again. . .

Other species

Like the true scarab beetles, the dung rollers, members of the Dynastinae, produce cheeping sounds like nestling birds by vibrating the edges of their wingcovers against microscopic ridges at the end of the abdomen. The behaviour of dynastids has been noted in the tropics because several species cause damage to native crops. The Indian rhinoceros beetle (*Oryctes rhinoceros*) is a pest of coconut plantations, especially in the Pacific Island of Samoa where it was introduced early this century. It is said that the beetle rams its great horn into a clump of leaves to gain anchorage while it tears at the plant tissues with its jaws. The 10-centimetre (4-inch) long atlas beetle (*Chalcosoma atlas*) also causes damage to coconuts by eating the young flower buds. Its great fat larva lives mainly in old trunks of coconut palms but will also bore into the roots of coffee plants causing considerable damage.

One should also mention the longhorn beetles of the family Cerambycidae whose larvae spend years boring through the wood of trees and whose adults have long and sensitive antennae. One of the largest of these is *Titanus giganteus* which, excluding the very long antennae, grows to over 13 centimetres (5 inches). This species was named by Linnaeus in 1778, not from a specimen but from a drawing he saw, the only insect depicted in a book of tropical birds. Until the 1950s, few, if any, living specimens were seen and high prices were paid by collectors for dead specimens that were found by the South American Indians, floating on the surface of the Amazon and its tributaries. In the same league as the *Titanus* is the 12-centimetre (4¾ inch) long *Xixuthrus heros* from the Fiji Islands. Both these longhorn beetles are of a dull brownish colour and lack the thick cuticle and armour plating of the goliath and rhinoceros beetles.

The largest British beetle is the stag beetle (*Lucanus cervus*) of the family Lucanidae. It grows over 7 centimetres (2¾ inches) long.

The Blue Whale

Mammoth of the deep

Right: The blue whale (*Balaenoptera musculus*) is the largest animal on earth, measuring up to 33 metres (110 feet) and weighing over 150 tonnes.

Although not noted for his sense of humour the Swedish scientist Linnaeus must have amused himself a good deal when he gave the Latin name *musculus* (little mouse) to the blue whale (*Balaenoptera musculus*), the largest living animal in the world. Most references cite this enormous marine mammal as being the largest animal ever to have inhabited the earth. However, in the world of nature records are continually being broken and the blue whale was toppled from its position in 1972 when the remains of a new species of dinosaur, later nicknamed 'Supersaurus', were discovered in Colorado. The massive size of the bones found indicate that this giant reptile must have tipped the scales at well over 200 tonnes, a good 50 tonnes heavier than the largest blue whale. Nevertheless in the land of the living the blue whale retains its title. The largest specimen of blue whale ever caught was a female towed ashore to a whaling station at South Georgia in the South Atlantic Ocean which measured over 33 metres (110 feet) – nearly one and a half times the length of a swimming pool! The dimensions of an average blue whale are a good deal smaller with a length of 26 metres (85 feet) and a weight of 106 tonnes. To give a better idea of the enormous bulk involved, this weight is equivalent to 4 brontosaurus dinosaurs, 11 elephants, 30 cows or 1600 men!

What is even more remarkable about this giant is its phenomenal rate of growth. It takes only eleven months for the whale to develop from the fertilized egg weighing only a few grams to the young whale which at birth measures 7 metres (23 feet) and weighs over 2 tonnes. During the weaning period which lasts seven months development is also rapid for the young whale doubles its birth length and attains a weight of 25 tonnes in its first year of life. This means it must gain around 90 kilograms (198 pounds) in weight every day! The young whale is fed by its mother on a highly nutritious food having the consistency of condensed milk and being up to four times more concentrated than cow's or goat's milk. Like most mammals the blue whale suckles its young – what is amazing in view of the extremely fast rate of growth is that the young whale appears to suckle for short periods of time, a few minutes at most, although it may feed up to fifty times a day. The milk is forcibly ejected into the youngster's mouth and it has been estimated that the young calf consumes 600 litres (130 gallons) of milk per day. Because of the great demands made on the female during pregnancy and the weaning period normally only a single calf is born every second year although cases of twins are known.

Apart from its great size the blue whale is distinguished from the other great whales by its overall dark slate-blue colour. The body is torpedo-like with a beak-like flattened head which accounts for a quarter of its body length. It also retains vestiges of mammalian hair – about forty small whiskers mainly concentrated on the tip of the lower jaw. The blue whale's underside has the appearance of corrugated rubber with long grooves running down from the lower jaw along the throat and chest. These grooves, 5 centimetres (2 inches) deep, are thought to give elasticity to the skin allowing the whale to increase the volume of its mouth cavity when swallowing; they probably also aid in streamlining when the animal is swimming.

In the open sea these great bulky bodies serve as islands of attachment for many different kinds of small invertebrate animals such as barnacles, sea pens and whale lice. The

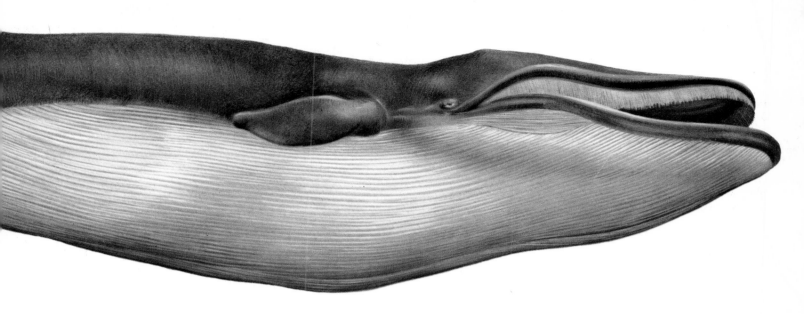

undersides of individuals, particularly those living in the North Pacific, are often encrusted with millions of tiny yellow plants called diatoms – this yellow coating has given them the common name of 'sulphur bottoms'.

The blue whale together with the other great whales such as the fin, right and humpback whales feeds by filtering small animals from the upper surface layers of the ocean. The filtering equipment consists of baleen plates which have been likened to vast string curtains which hang down from both sides of the upper jaw. The plates, 2 centimetres ($\frac{3}{4}$ inch) thick and 1.3 centimetres ($\frac{1}{2}$ inch) apart, have hairy fringes which form a coarse fibrous mat and function as a strainer, the water flowing freely through the holes while the food is held. As the whale closes its mouth its large flabby tongue presses down and squelches the food down onto the floor of the mouth forcing streams of water out between the baleen plates.

The staple diet of the blue whale and many other great whales consists of small crustaceans known as krill which resemble tiny shrimps. Krill abound in the upper layers of the polar seas during the spring and summer months, so much so that the waters may be stained reddish brown. The blue whales with their efficient filtering device scoop up these animals in vast numbers, probably taking up to 4 tonnes of them per day. By the end of the summer much of this food has been converted into a thick layer of blubber lying just beneath the skin. The blubber layer accounts for up to a third of the body weight of the whale. As the

short polar summer wanes and after a period of intense feeding the blue whales swim away from the pack ice to warmer equatorial waters. They travel singly or in pairs at a steady 8-10 knots although they are able to swim a lot faster if necessary. One tagged individual was recorded as travelling nearly 3200 kilometres (2000 miles) in just forty-seven days. The purpose of this migration is for breeding, for the calf when it is born has only a very thin layer of blubber which would be useless in insulating it against the cold waters of the Poles. During the period spent in the breeding grounds the adult whales feed very little if at all, using up their great store of body fat until by the time they make the return journey in the spring they are quite emaciated.

Monster Massacre

It was as late as 1967 when the populations of whales had been reduced to levels which made them no longer viable to hunt that the IWC called a ban by all its member countries on hunting the blue whale. Some nations such as the Japanese, who have always consumed a large quantity of whale meat, skirted the issue by setting up whaling companies in countries not affiliated to the IWC like Peru and Chile, and so the hunting of the blue whale continued, albeit on a very much reduced scale. Since its protection, the depleted populations of blue whales have shown some recovery, but there are still probably fewer than 10,000 of these mighty mammals throughout the great spread of the world's oceans.

The Marmoset

Monkey in miniature

Marmosets are active, agile arboreal monkeys inhabiting the dense tropical forests of the Amazon basin. Of the thirty-five known species the pygmy marmoset (*Cebuella pygmaea*) is the smallest, measuring just 30 centimetres (1 foot), over half of which comprises the tail. Unlike the majority of marmosets which sport a great variety of ear tufts, manes and whiskers, the pygmy marmoset has a smooth-furred rounded head and oblique 'mongoloid' eyes. The overall colour of the fur is brownish with a greenish cast, the hands and feet are yellow-orange and the tail indistinctly barred with black and brown. This species is an expert climber, scampering up and down and even backwards along branches by gripping firmly with hands and feet and making sharp jerking movements. When disturbed it spirals up a tree like a woodpecker. It communicates with a number of high-pitched 'chik-chik-chik' sounds. The pygmy marmoset was discovered by the German explorer Johann Baptist von Spix in 1823 in a forested area of the upper Amazon where Peru, Brazil and Columbia join borders. For years after its discovery it was thought to be the juvenile stage of the common marmoset (*Callithrix jacchus*).

Marmoset Species and Society

The common marmoset from eastern Brazil is the most familiar species, often kept as a pet or used as a laboratory animal to study primate behaviour. It is an extremely social animal that in the wild lives in large troops which move through the upper forest canopy. From studies on captive animals there appears to be a marked social hierarchy, and communication between members involves a large range of sounds and postures. Scent also plays an important role, secretions from the genital glands being used to mark branches and featuring in both courtship and threat displays. If a dominant male is approached by another it turns its back on him and raises its tail displaying its genitals. The other male approaches cautiously and may either sniff the tail or genital glands before retreating in a crouched submissive posture, uttering a peculiar mewing sound. Cohesion amongst the group is maintained by mutual grooming sessions though males are more often groomed than females. An animal that wishes to be groomed approaches its companion, looks directly at it and stretches out in an inviting posture. The groomer gets a good grip on a branch with its hind legs and meticulously proceeds to comb its companion's fur with its long-clawed nails, removing any parasitic insects and small objects trapped in the silky hair. During courtship a male will lovingly groom his mate, then the pair will walk after one another with a stiff-legged arch-backed gait. This culminates in the male approaching the female with a series of lip-smacking sounds interspersed with tongue flicking.

Marmosets show great attention to their young and individuals apart from the parents may play a part in rearing. As the young are born the male places them on his back and every two or three hours passes them back to the female for feeding, who when finished with them hands them back to her mate. After three weeks the young start exploring for

Below: The cotton-head tamarin (*Oedipomidas oedipus*) is confined to an area of coastal forest in Colombia. It is small but ferocious and will take a variety of animal food including lizards, mice and birds which are killed by a single bite to the skull.

themselves and accept solid food when they are four weeks old. Often the male will help the young with their first few mouthfuls of food by squeezing it through his fingers to make it more digestible. Marmosets communicate through a large repertoire of sounds which possibly includes an ultrasonic range undetectable by man. The alarm call of the common marmoset is a series of bird-like chirrups accompanied by head swivelling and short outbursts of 'gee-gee-gee'. This species is omnivorous, taking fruit, insects, spiders, lizards and small birds, which are flushed out from the dense cover by the mass movement of the troop. At night the troop 'encamps' in tree holes. The most predatory species is the black-tailed marmoset (*C. argentata*) which travels in groups of twelve or more among the shrub layer disturbing birds and small mammals killed with a quick bite on the head.

One of the most attractive of all marmosets is the golden lion marmoset (*Leontopithecus rosalia*) which has been described by one author as 'the most brightly coloured of all living mammals', being an intense shimmering golden yellow. Unfortunately this tiny mammal is now extremely rare. It was once found throughout the coastal forests of eastern Brazil, living mainly at 10–30 metres (approximately 30–100 feet) from the ground where dense vines, epiphytes and interlacing branches provided cover for them and their insect prey.

However, large-scale plantations of cocoa and rubber have eliminated most of the forest leaving isolated pockets such as the Poco das Antas reserve south-west of Rio de Janeiro which was established by the Brazilian government in 1973. Fewer than 200 golden lion marmosets are thought to survive in this 3000-hectare (approximately 7500-acre) forest. It appears to be their final stronghold.

Left: The common marmoset (*Callithrix jacchus*) lives a totally arboreal life in the dense forests of eastern Brazil. It is a highly social animal, communicating with other members of the group by a large range of calls and postures.

Left: The golden lion marmoset (*Leontopithecus rosalia*) is an endangered animal due to the destruction of its forest habitat.

The Giraffe

A living periscope

Far right: Standing its ground against attack, one male giraffe spreads its legs for balance whilst a rival swings its head towards its flanks. These sledge hammer blows continue until one giraffe suddenly retreats followed by the victorious male who holds his head up high in an arrogant gesture of dominance.

'An Abyssinian animal, taller than an elephant, but not so thick' was all Samuel Johnson had to say about the beast in his dictionary published in 1755. It is not surprising that his account was brief and to the point for at that time few Europeans had had the privilege of seeing a giraffe. It was not until the mid nineteenth century that a live specimen was brought to England by ship from Africa and it must have caused a sensation as it walked through the streets and suburbs of the metropolis on its way to the Zoological Gardens.

The giraffe is indeed remarkable, being the tallest animal on earth. The largest specimen on record is an adult bull shot in Kenya in 1934 which measured 5.87 metres (19 feet 3 inches) – over a metre taller than a London double-decker bus. Until 1901 when the okapi was discovered in the forests of the Congo, the giraffe was the sole representative of the family Giraffinae which is included with deer and antelope in the large mammalian order Artiodactyla. There is in fact only one species of giraffe (*Giraffa camelopardalis*) but because of variation in coat colour and pattern, and also because of the size and shape of the horns, nine distinct sub-species of giraffe are recognized, spread over the African continent south of the Sahara. The giraffe once had a more extended range – we know this from descriptions and illustrations in manuscripts and also from the numerous rock paintings where giraffe are depicted running with ostriches and lions, animals that also formed the faunal element of North Africa. Even during the latter part of the nineteenth century giraffe were found in north-west Africa especially near the Aïr massif in Niger, an area which not long ago supported a rich and varied vegetation; within the last hundred years the area has become progressively more arid, the once lush vegetation being covered by shifting sands.

Anatomy of a Giant

Its size, unusual proportions and the striking pattern of its coat make the giraffe one of the most fascinating of all large animals. The most striking feature is its neck which is both strong and flexible. Amazingly enough the neck is supported by the normal contingent of seven vertebrae but in the giraffe these are tube-like and extremely elongated, measuring nearly 30 centimetres (1 foot) in length. Like those of camels the vertebrae link together by a ball-and-socket type of articulation, the anterior end of one being rounded and fitting into the concave posterior end of the next one. This type of joint enables the giraffe to stretch and bend and curl its neck in all positions when feeding and grooming itself.

The problem of how a giraffe keeps its blood pressure steady has always intrigued zoologists because the head can at one moment be 3 metres (10 feet) above the level of the heart and then suddenly drop to 2 metres (6 feet 6 inches) below. From dissections and by measuring the blood pressure in various parts of the circulatory system of living animals it has been found that although there is nothing remarkable about the heart itself, the large veins have valves all along their length preventing the backflow and pooling of blood into the lower parts. Also, the two blood vessels (the carotid and external carotid artery) which feed the brain with blood from the heart divide before reaching the brain into a tight network of small vessels called the *rete mirabile*. The elastic walls of these vessels compensate for the movements of the head; when it is lowered they take up any excess blood preventing the head from being flooded and when the giraffe returns to the upright position enough blood remains in the *rete mirabile* to feed the brain without it being temporarily deprived.

The head itself is small in proportion to the rest of the body and functions like a periscope above and between the trees. The large eyes fringed with magnificent curly lashes have acute vision, able to spot predators or other giraffes clearly at long distance. Although they are not vulnerable to many animals, giraffe are almost constantly on the alert and sleep for short periods of a few minutes at a time. The giraffe's horns are curious appendages which

forty species in southern Africa. When browsing the giraffe uncurls its long 40-centimetre (16-inch) tongue, wraps it around a thorny branch and draws it gently between its extended lips. It then closes its mouth and combs leaves, thorns and twigs into its mouth. Some acacias harbour colonies of large black stinging ants which feed on the nectar produced at the base of the leaf stalks, and live in plant galls. It was believed that the giraffe would continue munching its way through an acacia bush without any apparent harm from the ants, but close observation has shown that once it has disturbed the ants it moves quickly on to an adjacent patch of leaves and seldom lingers long enough to provoke an attack. It is probably by accident that the giraffe was recorded as eating a weaver bird nest and its contents but giraffe have been known to eat the meat of a dead animal carcase for its salt and moisture content rather than for its flesh.

A giraffe chews its cud like a cow and has the typical four-chambered stomach associated with ruminants which consists of the rumen, reticulum, omasum and abomasum. Once the food is ground up by the broad teeth it passes into the rumen and reticulum where it is softened with gastric juices and partly fermented. When the animal is resting it chews its cud – round balls of partly digested food about the size of a man's fist are passed back to the mouth where each bolus of food is chewed for half a minute or so before being passed back down into the omasum and abomasum.

Behaviour

Giraffe society is loosely structured. They live either as solitary individuals or in small shifting herds of from three to thirty in open, wooded savannah and thorn bush country. Despite the lack of territorial behaviour bulls give way to each other according to rank, the strongest bull in a herd tilting its head at an angle to show its dominance among other males. Prior to the breeding season males fight each other using their heads as sledgehammers to slam each other with tremendous sweeps of their massive necks. As the fighting becomes more intense the blows get more violent and both males stand with their forelegs well apart in order to maintain their balance. There have been cases of fights to the death but in most cases one of the males gives way, falls to his feet or else the fighting stops abruptly.

Above: The giraffe's long flexible neck allows it to reach vegetation up to 6 metres (20 feet) from the ground. When feeding, the long prehensile tongue pulls a small branch down towards the mouth where the leaves are stripped off by the large fleshy lips.

arise not merely as outgrowths of the skull, but originate from separate centres of ossification. At birth they are soft cartilaginous knobs fringed with short tassels of hair, but they soon harden and become fused to the skull. Apart from the two main horns there is a horny bump on the middle of the forehead and two smaller horns at the back of the head which are thought to serve as points of attachment for the tendons, ligaments and muscles of the neck and back. Even for its small size the skull is light due to numerous large air sinuses.

Giraffe feed on a variety of trees and bushes especially the acacia of which there are over

Once a male singles out a receptive female he nuzzles her flanks, rubs his head against her haunches and nibbles her tail. These obvious caresses stimulate the female to urinate which forms a vital part of the courtship ceremony. The male lowers his head to catch the urine on his extended tongue and then raises his head and with mouth closed unfurls his large fleshy lips. This 'flehmen' or urine tasting behaviour is thought to test the hormone content of the female's urine. If this is satisfactory, both sexes rub and entwine necks prior to mating when, due to the male's weight, the female is forced to walk a few paces forwards. As one would imagine with an animal as large as a giraffe there is a long gestation period of fifteen months and when the 2-metre (6-foot 6-inch) youngster is born it comes into the world with a mighty bump which apparently does it no harm. Like all mammals the female suckles her young but a kind of 'kindergarten' structure exists among giraffes, so that a few months after birth the young giraffe is often looked after by another female.

Although maternal behaviour is not as well developed as in other animals like the rhinoceros which keeps a constant vigil on her bulky youngster, a female giraffe will defend her young by lashing out with her legs. At the approach of danger the young giraffe springs between its mother's forelegs broadside on so as to keep its head away from the predator. Both lions and leopards have been found mortally wounded after a blow from a giraffe's back legs, but this does not deter them from attacking a giraffe especially when hunting in pairs. A couple of lions will manoeuvre themselves downwind of a group of giraffe then slowly crawling on their bellies will approach slowly and silently until they are within striking range. While the giraffe is momentarily preoccupied with feeding, one of the lions will spring onto the giraffe's back and attempt to inflict a disabling neck bite.

Below: When it splays its legs wide to drink at a waterhole a giraffe is especially vulnerable to predators.

Above: A group of reticulated giraffes ambling along in single file. This is one of the most handsomely marked sub-species, which is confined to the open bush of Kenya, Ethiopia and Somalia.

A group of giraffes lolloping across the savannah is one of the most characteristic picture-book images of Africa. Because of its unusual proportions locomotion differs from other even-toed herbivores. During the peculiar pacing walk the two legs on one side of the body move forward at nearly the same time. When pursued or disturbed the walk develops into a gallop. With the tail curled over the back the hind limbs push off and are thrown well forward of the forelimbs. In this way the animal takes enormous strides covering large distances even though the pace of the gallop does not appear very fast. When walking and

running added momentum comes from the head and neck which move backwards and forwards twice during each stride. Giraffe can maintain a steady speed of 56 kilometres per hour (35 m.p.h.) for two hours though after this they become exhausted. Because of their high steady speed Arab hunters used to use relays of horses to bring down a single animal.

Protection

The large bodies of giraffes have provided the peoples of Africa with a good supply of raw materials. The flesh of the young and of the females is said to be especially tender, which has led to giraffes being farmed in some places for their meat. The hide has been made into hunting shields, sandals and amulets used as a charm against lions. But the most valuable part of the giraffe has always been its tail, valued by tribal chiefs as a magnificent tail switch. The individual hairs are used by Masai women as thread to sew beads onto their clothes. Despite the predation of the giraffe by the aboriginal inhabitants of Africa the population maintained itself. Once the white man arrived the situation changed dramatically – the Boers in South Africa were especially destructive, running down and shooting hundreds of giraffe in a single day. The spread of human population and more intensive agriculture have all contributed to the decrease in the giraffe's range and one sub-species, Rothschild's giraffe, is especially vulnerable.

Today, giraffes are a valuable asset to many African countries in which the revenue accruing from the tourist industry may account for a good deal of the country's income, and so they are protected in many National Parks and nature reserves. Kenya is especially fortunate as three giraffe sub-species live within its boundaries. Giraffe can even be seen in Nairobi National Park, just 8 kilometres (5 miles) from the centre of Nairobi, browsing peacefully in company with rhino, zebra and a multitude of smaller antelope.

Above: Though a young giraffe can nibble grass a week after birth, it is weaned for a period of up to ten months.

Relative sizes of man and giraffe

THE RELATIVE SIZES OF MAMMALS

Name	Height or Length (adult male)	Weight (adult male)
Man *Homo sapiens*	Total height 1·8m	70 kg
Blue whale *Balaeonoptera musculus*	Head and body 22–30m Females average 0·5m longer than males	100,000–130,000 kg
Etruscan shrew *Suncus etruscus*	Head and body 35–48mm Tail 25–30mm	1·5–2·5 g
African elephant *Loxodonta africana*	Length of head, body and trunk 6–7·5m Tail 1–1·3m Hieght at shoulder 3–3·5m	5000–6500 kg
Giraffe *Giraffe camelopardis*	Total height 4·8–5·2m (including 'false horns')	1000–1400 kg
Tiger *Panthera tigris.*	Head and body 1·8–2·8m Tail 0·9m	227–272 kg
Gorilla *Gorilla gorilla*	Total height 1·6–1·75m	150–200 kg
Pygmy marmoset *Cebuella pygmaea*	Head and body 130–144mm Tail 190–210mm	70–150 g
Red kangaroo *Macropus rufus*	Head and body 1·4–1·6m Tail 0·8–1·0m	70–75 kg
Flat-skulled marsupial mouse *Planigale subtilissima*	Head and body 44–50mm Tail 50–56mm	4–4·8 g
Red deer *Cervus elephas*	Height at withers 1·1–1·4m	85 kg (moorland) 150 kg (English woodland)
Pygmy shrew *Sorex minutus*	Head and body 40–60mm Tail 30–40mm	2·5–5·0 g

Weight at birth	Points of interest
2·5–6 kg Usually single young	Man, like many of the other animals in this table, varies considerably in height. The measurements given should provide a scale to judge the variety of sizes and weights found in the class Mammalia.
2,000 kg Usually single young	The largest animal living or known to have lived, blue whales feed on the shrimp-like crustaceans known as krill. One specimen measuring 26 metres long contained 2,000 kg of krill, extimated to be over 5 million individuals.
No record	The smallest living mammal (though some bats may weigh less). Feeds on spiders, grasshoppers and other insects.
90–120 kg Usually single young	The largest land-living mammal. Feeds on coarse plant material of low food value, and has to spend up to 18 hours a day browsing on trees and shrubs to obtain enough nourishment.
70 kg Usually single young	The tallest of living mammals, the giraffe feeds mainly on leaves and branches of thorny trees. In areas where many feed there will be a 'browse line' at about 5 metres height of the trees, above which even giraffes cannot graze.
1·5 kg 3–4 cubs	The largest land-living carnivore, both longer and heavier than the lion. Tigers are forest dwellers and feed mainly on deer, wild boar and antelopes.
2 kg Usually single young	The largest of the primates, the group which includes man. Female gorillas are considerably smaller than males. The inactive life of a zoo causes overweight in many individuals in captivity.
No record Usually 2 young	The smallest of the monkey family and arguably the smallest primate. Recent observation has revealed these marmosets feed mainly on the sap of various and shrubs, though they also eat fruit and leaves.
0·75 g	The largest of pouched mammals or marsupials, kangaroos also have the smallest young in relation to the adult. They are born at an extremely immature stage, with the hind limbs only buds, and pull themselves to the safety of the mother's pouch.
No record 4–12 young found in pouch	The smallest of marsupials, these tiny pouched mice feed on grasshoppers as large as themselves. A captive specimen of the very closely related *P. ingrami* ate 6–8 5cm long grasshoppers each day.
5–10 kg Usually single young	The largest of British land mammals, red deer vary greatly in size depending on their habitat. Scottish moorland stags are half the weight of English woodland dwellers, while deer from European forests may weigh up to 255 kg.
0·25 g 2–8 young born	The smallest British mammal. Its very high rate of activity makes the pygmy shrew spend most of its time hunting for the woodlice spiders and beetles that are its food; research on captive specimens showed they ate over 3 times their own weight daily.

FEROCITY
—AND—
VENOM

The formidable gape of the
snapping turtle *(Chelhydra serpentina)*,
a North American species which grows up
to 13.5 kilograms (30 pounds).

Poisonous Snakes

Original enemy of man

Snake-bites account for over 30,000 human deaths each year, more than all those caused by man-eating tigers, leopards, killer sharks, crocodiles and the odd rogue elephant. Of the 3000 known snake species at least 400 are regarded as dangerous. The one with the most potent venom is a sea snake (*Hydrophis belcheri*) which grows up to a metre (3 feet 3 inches) long and lives in the shallow Timor Sea some 320 kilometres (about 200 miles) north-west of Australia's Northern Territory. The venom of this snake is incredibly toxic – at least a hundred times deadlier than the poison of any terrestrial snake including the cobra, taipan and viper. Fortunately the sea snake is not aggressive, has a small head, relatively weak jaws and is seldom the cause of a human fatality. Its normal prey are small eels which it hunts by slithering into the narrow crevices and burrows in which they hide.

Various types of snake poison affect the body in different ways, but generally they are either neurotoxic in action, affecting the nervous system, or haemolytic, destroying the lining of the blood vessels and causing a clumping together of the red blood cells. Very often a snake's venom will contain a mixture of these elements plus other poisons that attack specifically the white blood cells and also the muscle tissue. The venom of cobras and sea snakes tends to be neurotoxic in effect whereas vipers' venom is haemolytic. In all cases the venom is produced by modified salivary glands, and has evolved to subdue the prey, and to start the process of digestion.

Cobras and Mambas

One of the most impressive poisonous snakes, on account of its large size, is the king cobra or hamadryad which grows to over 5 metres (about 16½ feet), and when about to strike typically rears up a quarter or one-third of its body length off the ground. This magnificent sleek reptile is a snake-killer, taking other cobras, kraits and vipers as well as a range of non-poisonous snakes. It is also responsible for a large number of human fatalities in India, Burma and Thailand where the largest speci-mens are found. The famous explorer Colonel Frank Wallace recorded in detail the symptoms of a cobra bite:

The earliest constitutional symptom is a feeling of intoxication. Later the patient feels his weakness insidiously creeping upon him, till uncertain of maintaining upright posture, he voluntarily reclines. His paralysis begins in his legs, mounts to the trunk, and finally affects the head which droops. Synchronizing with this drooping of the head a drooping of the eyelids may be noticed, and simultaneously the muscles of the lips, the tongue and throat become gradually paralysed. As a result the lower lip falls away from the teeth, allowing the saliva to dribble from the mouth. Speech becomes increasingly difficult till the subject, unable to control his lingual and labial muscles, attempts by signs to communicate to those around him . . . breathing becomes embarrassed, laboured and finally impossible so that fluids taken in through the mouth are apt to be regurgitated through the nose. Nausea and vomiting are frequent. A convulsion often heralds the cessation of respiration, but the heart goes on beating for a minute or two longer. Consciousness is retained until the end . . . Such are the effects of the paralysing influence on the nerve cord and the brain, which may cause death in from 1½ to 6 hours.

The king cobra (*Ophiophagus hannah*) can and has killed elephants, biting them in the soft skin between the toes, and even these huge animals take only three hours to die. Apart from its size and potent venom this species is unusual in that the female constructs a nest for her young. Bamboo leaves and dried stems are often chosen for the nest, the female gathering them in her body coils.

The most familiar species of cobra is the Indian cobra (*Naja naja*) often photographed being 'charmed' out of a straw container. According to the author and traveller James Clarke this snake is the biggest killer in India

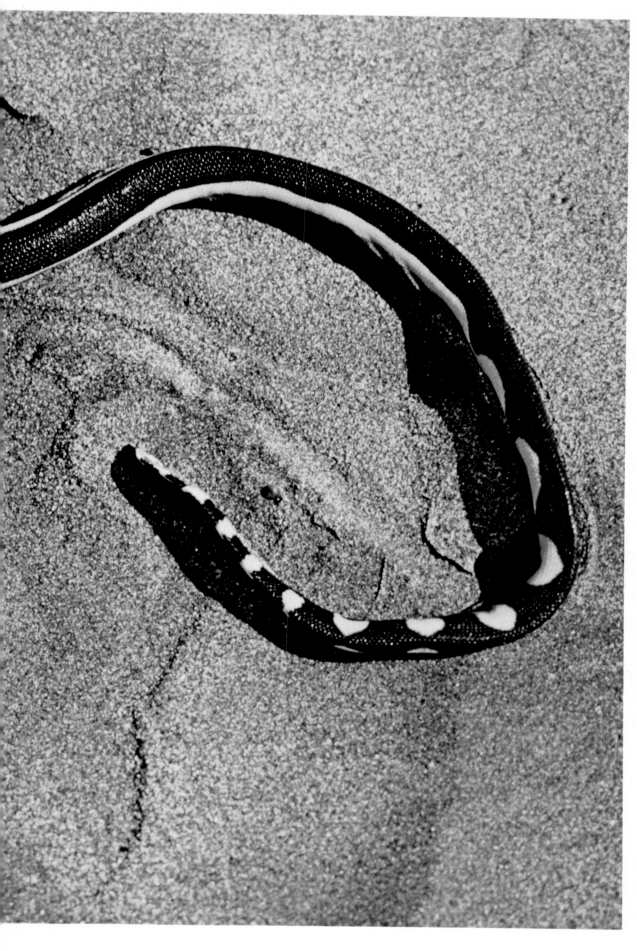

Left: The pelagic sea snake (*Pelamis platurus*) ranges across the Pacific where it feeds on small fish in the surface waters. All fifty known species of sea snake are extremely poisonous. The venom of the small-headed sea snakes of the genus *Hydrophis* is especially virulent.

partly because it is most common and also because it is found around human dwellings. India and the Far East have many species of cobra but many more are found in Africa, although some lack the characteristic inflatable hoods. Two of the most common African cobras are the asp (of Cleopatra fame) otherwise known as the Egyptian cobra (*Naja haje*), and the Cape cobra (*N. nivea*) regarded by many as the most venomous species. However, the most dramatic in their method of hunting are the spitting cobras of which the black-necked cobra (*N. nigricollis*) has the most notorious reputation. This snake stuns its prey by shooting out twin jets of venom, aiming for the eyes. At a range of 2.5 metres (about 8 feet) its aim is deadly accurate, blinding its victim before pumping even more venom into it with its fangs. A man hit by such a snake can suffer permanent blindness if the venom is not washed away within a short space of time.

In Africa mambas, closely related to cobras, have an even more sinister reputation; they are certainly far more aggressive snakes. The black mamba (actually a dull-greyish colour) is a thin 3-metre (10-foot) long, sleek terrestrial snake which is credited with being the fastest snake in the world although in fact it can only move at 11 kilometres per hour (7 m.p.h.) over short distances. This snake (*Dendroaspis polylepis*) often hunts in long grass and has the formidable reputation of striking at anything that moves in its path. It can also strike a man high up on his body as it has the ability to rear up over forty per cent of its body length. The brighter-coloured green mamba is a slender arboreal snake which is equally dangerous, striking repeatedly.

Two other snakes which have the same fixed fang arrangement of venom apparatus as the cobras are worthy of mention – these are the taipan and tiger snake of Australia. Both snakes have claimed a number of victims, the taipan (*Oxyuranus scutellatus*) having the worst reputation although it is only found in the humid north-east of the country whereas the tiger snake (*Notechis scutatus*) is not uncommon in and around the suburbs of Sydney. Fortunately antivenins have been produced for both species and fatalities from snake-bite in Australia are now quite rare.

Vipers and Rattlers

The other main grouping of venomous snakes includes the Old World vipers and the rattle-

snakes of the Americas. The word viper conveys the idea of a small evil snake with a painful if not deadly bite. This may well apply to the common viper (*Vipera berus*) but there are many larger and more dangerous species.

Vipers have an extremely wide distribution and include some of the snake species which are most tolerant of cold. For example the common viper or adder in Scandinavia ranges well inside the Arctic Circle. This viper's bite is seldom fatal but other European species particularly *Vipera ammodytes* are classed as extremely dangerous. The most deadly of all is the insignificant metre-long (3 feet 3 inches) saw-scaled viper (*Echis carinatus*) found from North Africa through the Middle East into India and as far south as Sri Lanka. The lightning speed of its attack, its nervous aggressiveness and especially potent venom make it one of the most feared of snakes and it is, unfortunately, one that is quite frequently encountered. Before attacking it typically coils its body into a moving figure of eight, the rubbing of its skin producing a sharp hissing sound. Most casualties from this snake are bitten at night and most often on the foot or ankle.

Whereas the bite of a cobra is slow-acting and has an initial soporific effect, that of a viper is extremely fast-acting and usually very pain-

Above: The horned viper (*Vipera ammodytes*) is a small but dangerous snake of southern Europe. A bite from any viper is painful – the poison destroys the blood vessels causes skin discoloration, internal haemorrhages and breakdown of muscle tissue.

ful. After a viper has struck there is an acute burning sensation, the area around the bite swells up to alarming proportions and becomes extremely sensitive to touch. This swelling gradually spreads over the body and the initial site of the bite and the surrounding tissue become discoloured due to local haemorrhage. The flesh may look blue, green, purple or black with livid blotches, dark red spots, blisters and streaks; the effect is disgusting, as though the body was literally being eaten away. This change of skin colour has given rise to the widespread superstition that a human victim takes on the colours and patterning of the snake that bit him.

The sixty species of true vipers are spread over a variety of habitats from open moorland and tundra to the most arid deserts. They are also plentiful in forest areas of the tropics where several species show beautiful cryptic coloration. One of these, the Gaboon viper (*Bitis gabonica*), holds the record for having the longest fangs of any snake – up to 10 centimetres (4 inches) and also on average contains the greatest quantity of venom, over 1000 milligrams (0.035 ounces) per individual. Fortunately this snake is fat and sluggish, with a rather retiring nature unless accidentally trodden on, whereupon it hisses and growls and once having sunk in its fangs is reluctant to let go of its victim! Other dangerous vipers include the smaller African puff adder (*Bitis arietans*) which accounts for the most number

of snake-bite cases in Africa, and Russell's viper (*Vipera russelli*) found throughout India and the Far East.

On the other side of the world the true vipers are replaced by the pit-vipers which include rattlesnakes. Pit-vipers are so-called because they possess a pair of openings between the eyes (called pits) that are heat-sensitive. These special sense organs help the snake not only to detect prey in the dark but also to keep in contact with a bitten animal which has struggled some distance into the bush.

America's south-west is literally crawling with rattlesnakes, the state of Arizona having a record seventeen species. Rattlesnakes and vipers show the most highly developed poison mechanism – the fangs are large and when not in use are folded on the roof of the mouth. When however the snake makes a strike the mouth is opened wide and the fangs swing down through an arc of approximately ninety degrees. The curved fangs are dagger sharp and housed in a sheath which is withdrawn when the snake bites into its prey. Of the many rattlers the western diamondback (*Crotalus atrox*) causes most fatalities, while the Mojave rattlesnake (*C. scutulatus*) and the eastern diamondback (*C. adamanteus*) contain the most virulent poisons. Although rattlers are a common ingredient of the Western film, they also range far north into Canada, including the prairie lands of Alberta and the coniferous forest zone of Ontario. The various species

Right: The Gaboon viper (*Bitis gabonica*) of Central Africa is one of the largest venomous snakes, reaching a length of nearly 2 metres (6½ feet) and a weight of 8 kilograms (18 pounds). It is also notorious in having the longest fangs of any snake, up to 10 centimetres (4 inches) long, and a high venom yield. Fortunately it is slow-moving and does not strike readily.

Left: A puff adder
(*Bitis arietans*) opens
its large mouth as a
threat. When
excited, this species
will also puff its body
up with air – in this
way small individuals
can double their
thickness and make
themselves appear
more formidable to
would-be predators.

Right: The most advanced poison mechanism is shown by vipers and rattlesnakes. The large fangs are set in bones which can be rotated so that the fangs lie flat or are swung out when ready to strike. Once the fangs make contact, the venom is forcibly pumped into the prey.

Left: The timber rattlesnake (*Crotalus horridus*), one of several deadly North American species.

Left: The copperhead (*Agkistrodon contortrix*) is responsible for many cases of snake-bite but few fatalities.

at rest

upper jaw   quadrate

maxilla   fang   lower jaw

striking

swallowing

have their own venomous concoctions combining both neurotoxic and haemolytic poisons in differing proportions.

A rattlesnake's rattle is made up of interlocking segments of horny keratin which vibrate against one another when violently shaken. Although some fanciful theories have been put forward as to the rattle's function, it appears to be simply a rather efficient defence mechanism which in many cases saves the snake from expending the energy of an attack and from wasting its reserves of venom. Apart from rattlesnakes North America has two other potentially dangerous snakes, the copperhead (*Agkistrodon contortrix*) and the cottonmouth or water moccasin (*A. piscivorus*). Of these the copperhead probably bites more people than any other snake in North America although its bite is seldom fatal, while the less frequently encountered cottonmouth is very dangerous, its venom being equivalent in strength to that of some of the most virulent of rattlers.

Central and South America are infested with pit-vipers both terrestrial and arboreal. Of these the bushmaster (*Lachesis muta*) is probably the largest, with the fer-de-lance (*Bothrops lanceolatus*) and jaracara (*B. jaracara*) being equally dangerous. There is also the jumping viper (*B. nummifer*) which can launch itself off a branch or even leap up from the ground to attack – this snake must be everyone's idea of a jungle nightmare!

The list of dangerous snakes seems endless and there are probably very many more species awaiting discovery in the remoter forest areas. Whatever their size or venomousness, however, most snakes are not dangerous unless provoked. If one has the misfortune to be bitten the rule is to stay as calm as possible since panic only aggravates the condition, making the heart beat faster and distributing the poison more quickly round the body. Antivenins exist for the more common species.

The Polar Bear

King of the icy wastes

The polar bear (*Thalarctos maritimus*) is not only one of the largest bears in the world, it is also the most carnivorous, a true hunter spending most of its adult life alone, wandering the pack ice in search of seals. The polar bear is a truly formidable beast – only the Kodiak bear of Alaska matches it in size. An adult male bear weighs over half a tonne and when reared up on its hind legs stands around 3.4 metres tall (over 11 feet). Apart from its sheer size it is beautifully proportioned for locomotion over ice and snow. The paws of the forefeet serve as massive snow shoes and the short sharp claws are ideal instruments for gripping slippery and often sheer ice faces. Its teeth are long and pointed for tearing the skin and blubber off seals and whale carcases.

It is no wonder that the early explorers were terrified of the great white bear whose appearance was so sudden out of the huge expanse of whiteness. One of the earliest accounts of a polar bear attack comes from the journals of the Danish-Russian explorer Bering who pioneered Arctic exploration:

> The bear at the first falling of man, bit his head in sunder and sucked out the blood, wherewith the rest of the men . . . ran presently thither, either to save the man or else drive the bear from the dead body; and having charged their pieces and bent their pikes, set upon her, that still was devouring the man, but she perceiving them to come towards her, fiercely and cruelly ran at them and gat another of them out from the company, which she tore to pieces wherewith all the rest ran away.

One can imagine the panic and confusion that this particular bear caused, but since 1596 when the report was written there have been many more accounts of the stealth and cunning of a hunting polar bear. These bears are also on record as having tried to smash down igloos to get at women and children shut up inside while the men were away on a hunting expedition. In Eskimo mythology the polar bear is considered a supreme being who hears and understands all that man talks about and who after death can transform itself into a human form and back again into the body of a bear – a kind of Arctic werewolf. There are many taboos connected with the hunting of polar bears, for example, when one is killed, depending on its sex, various hunting implements together with a piece of meat or blubber are stored with the skin, providing the soul of the bear with nourishment. Most attacks on man are by adult males who out of sheer hunger will kill and eat anything they can overcome. They do not relish human flesh, however, but much prefer the skin and blubber of seals.

Skirmishes with man are incidental in the life of the polar bear which spends much of its adult life as a solitary nomad wandering along the coastal ice floes and pack ice of the Arctic. The moving streams of ice which travel in a clockwise direction around the Pole serve as a moving highway for bears in their search for food. Whilst its movements are governed to some extent by the seasons and the slow breaking up and formation of pack ice, the polar bear's migrations are not merely passive meanderings but are geared to finding the seals which serve as its principal prey.

The Great White Hunter

In the early spring when the bears emerge from their dens they prey mainly on young ribbon seals born in tunnels in the snow above the mother's breathing hole in the ice. These burrows, termed 'aglos', are often covered several metres deep in snow, yet a polar bear with its keen sense of smell can detect if a burrow is occupied and probably the state of development of the young seal. After inspecting several aglos a bear will choose one and quickly scrape off the outer covering of ice. It then pounds down on the dome with the full force of its body and smashes up the chamber, holding down the pup before it escapes.

As the season progresses the young seals remain vulnerable, for even though they are fast swimmers they have to come to the surface to breathe more frequently than

adults. Also, being inexperienced in the ways of the polar bear they often loll about near ice hummocks which afford ideal cover for the predator. When hunting these juveniles on the ice a bear crawls slowly forward a bit at a time commando-style with the hind legs trailing, then, when within a few metres of its quarry it springs to the attack, cutting off the seal's escape route to the water.

As the summer sun becomes stronger the pack ice breaks up and the bears take to hunting adult seals around leads in the ice. This type of hunt necessitates even greater stealth, for adult seals are by nature extremely wary animals, sleeping for only a few minutes at a time before swivelling their heads around to investigate their surroundings. Having spotted a seal on the ice a bear silently lowers itself into the water hind feet first and using its broad paws as paddles swims slowly nearer with only the tip of its long nose causing a faint ripple on the water. Sometimes a bear will use a block of ice for cover to rise up out of the water and gauge the distance to the quarry. Then, diving under the ice, the bear will position itself just below the resting seal and with a surprise attack lunge upwards, breaking the ice and knocking the seal into the water where it is killed with a mighty blow of the forepaw.

Another method is to surprise a seal at its blow hole where the bear takes advantage of the few seconds when as the seal exhales, it cannot hear the approaching danger. The seal is hit on the muzzle and with one sweep of the bear's forelimb is scooped out onto the ice. The fact that even a bearded seal, which may weigh almost as much as its attacker, is dealt with in this way is an indication of the power and co-ordinated strength of the polar bear.

Mating

Polar bears only come together in the spring to mate. Males will spar viciously to gain a female and once the bears pair off they may stay together for a few weeks during courtship and prior to mating. Once the female is fertilized the male returns once more to a life of isolation. Polar bears exhibit a phenomenon called delayed implantation, the fertilized egg remaining in a state of arrested development for some months, and it is not until the autumn that the female bear searches out a suitable site in which to excavate a den. This consists of a tunnel which leads into one or more roomy chambers. In one of these chambers the bear rests up in preparation for giving birth – snowdrifts soon block the entrance and cover the tracks. Once the female is entrenched in her den she will not feed for at least four months, but survives by living off her thick reserve of blubber under the skin, and drinking melted snow.

The two young, born at the end of December or early in January, are naked and about the size of small rats when born, but they quickly put on weight so that by the time the den is opened up in the spring they weigh around 10 kilograms (22 pounds), have thick fur and are amazingly strong. For the next fifteen months or so the young remain with the mother who is a most attentive parent, teaching them how to deal with the most difficult and often treacherous terrain and most important of all how to hunt seals and become independent. During the period the cubs remain with her she avoids all contact with other bears who are potential threats to her cubs – cannibalism is not uncommon. As the trio of bears trek across the ice floes the mother will help the young forward by nudging them with her snout or alternatively giving them rides, one at a time, on her back.

Sparing the Bear

Though the polar bear is an expert hunter it has little defence against those who hunt it from helicopters using sophisticated long-range rifles. Thousands of bears have been shot in this way, ending up as big-game trophies or as foot warmers on the floor of some luxurious apartment. It was because of the increase in hunting which by 1970 had reduced the total population to around 20,000 bears that conservationists rallied round to protect them. In 1973 the International Union for the Conservation of Nature (IUCN) set up a conference specifically aimed at the protection of the polar bear throughout its Arctic range – the agreement proposed has now been ratified by all those countries with territories bordering the Arctic Circle. It is now prohibited to hunt or take a polar bear except for *bona fide* scientific purposes or by local peoples like the Eskimos. Since the agreement fewer pelts have reached the European market, and, although there is some poaching the survival of the great white bear seems secure.

Right: A female polar bear followed by her cubs swims away from shore. Although a powerful swimmer the bear is not fast enough to hunt seals in the water; instead it relies on stealth and suprise attack mainly on the pack ice.

The Killer Shark

Restless cruiser of the seas

No animal is more feared than the shark. Few people see the ominous triangle of its dorsal fin cut through tropical seas without a thrill of fear, perhaps mixed with wonder at the perfection of the outline beneath. Sharks are of many shapes and sizes, but the large ones that cruise the deep oceans have a highly evolved hydrodynamic form combined with the sensory and killing attributes of one of the world's most beautifully functional predators.

Legends of their ferocity are legion, but none of them is more impressive than the truth of their mysterious frenzy when feeding or than the single-mindedness with which they will continue to tear at their prey even when mortally injured themselves. The increasing appreciation of these little understood creatures only extends the wonder one feels for something that so precisely fits its purpose. For centuries misconceptions about sharks have clouded our picture of their place in the natural order. It is still possible to find people who believe that the shark, because of its poor senses, is led to its prey by the pilot fishes that scavenge along its flanks: but now we know a little more – though still far from everything – about the shark's perception of its surroundings, the animal appears to be even more wonderful. The clever little pilot fish are now known to play a much more modest role in the shark's life than the fishermen's story told.

Sharks have been successful in making homes for themselves in the waters of the great oceans, the shallows of its reefs and shorelines and even in the fresh waters of some lakes and rivers. One species has even migrated so far from what seems to be the natural marine habitat of sharks as to appear in lakes 60 metres (about 200 feet) above sea level in Papua-New Guinea. Worshipping Hindus sometimes suffer the attentions of the Ganges shark (*Carcharhinus gangeticus*) which, although small by shark standards have killed several pilgrims to its sacred waters. Of the 250 to 350 species of sharks, grouped into about twenty families, the most feared by humans belong to a division that includes twelve of these families: these are the Galeiformes or 'shark forms'. Among its members is the most dreaded of all, the great white shark (*Carcharodon carcharias*), which is the largest of all those known to attack man. Its still larger relatives, the whale shark (*Rhincodon typus*) and the basking shark (*Cetorhinus maximus*) – the two largest fishes in the world – peacefully browse the migrating plankton and shoals of small fish. Their enormous jaws are toothless, equipped only with masses of tissue that act as filters for their tiny but myriad prey.

The mackerel sharks (Isuridae) of the division Galeiformes are the all-powerful wolves of the deep oceans. They include the great white, the mako (*Isurus paucus*), porbeagle (*Lamna ditropis*) and thresher sharks (*Alopius volpinus*), all known to attack human swimmers. The requiem shark, whose name announces its sinister reputation, is of the Charcharhidae family, notorious enemy of careless bathers, who also enjoy the attentions of its close relatives, the tiger shark (*Galeocerdo cuvieri*), the white tipped reef shark (*Carcharhinus longimanus*) and the great blue shark (*Prionace glauca*).

All of the mackerel and requiem sharks are exceptional swimmers. The slight flattening of their heads and the round sections of their bodies may contribute a little to their speed through the water. Sharks are not particularly fast over long distances. Like those arch-predators of land animals, the tiger and lion, they prefer to lie in wait for their prey and sprint after them when they wander into range. It is difficult to measure the speed of a shark in short bursts as it charges its prey, but a small mackerel shark of only 60 centimetres (2 feet) in length produced bursts of 70 kilometres per hour (43 m.p.h.): a larger fish would swim substantially faster.

The shark is finely evolved for this kind of swimming. Its fins are constantly erect and cannot fold in towards its body. The pectoral fins spread stiffly to the sides like small wings to give the fish 'lift'; an important effect to a fish that has no swim bladder. Most fishes release gases into a sac which controls their buoyancy, but the shark's pectoral fins combine with its forward motion to give it the 'lift' needed to prevent the fish from sinking. The shark's

Right: The great white shark (*Carcharodon carcharias*) is the most dangerous and the largest of all the predatory sharks, attaining a length of up to 6 metres (20 feet). It lives mainly in warm or temperate waters and will tackle prey as large as bull sealions weighing up to half a tonne.

specific gravity, 1.06 – 1.09, is somewhat greater than that of seawater (1.026), but the shark sometimes reduces its density by accumulating oils of low specific gravity in its liver. The two lobes of the liver in some sharks occupy most of the length of the body cavity, and the oils supply an energy food for the animal when it has fasted for a period, such as when a male is preparing to mate or when a female is gravid. The amount of oil changes from season to season and with the shark's condition, but in most cases it accounts for about 5–15 per cent of its body weight. In exceptional circumstances the liver may be as much as 20–25 per cent of the shark's bulk, and half of it will be oil. Before the days when vitamin A was manufactured synthetically, the shark's liver oils caused many to be slaughtered for this precious product.

The angle of the shark's caudal fin (the upper part of its tail) and its crescent shape are ideal for swimming at speed. The fulcrum of its side-to-side swimming action lies well forward, in the nuchal region just behind its head, so that the leverage of the sculling movement of the rest of its body gives maximum drive. In those sharks, especially the thresher,

where the caudal fin is extremely developed, the sweeping action tends to drive the fish downwards, but the 'lift' from the pectoral fins counteracts this and converts the impulse into a strong forward drive. The three vertical fins act as stabilizers, holding the shark in its natural posture. The hammerhead shark (*Sphyrna zygaena*) shows the most extreme form of this planing effect produced by the pectoral fins. The hammerhead's extraordinary, flattened and laterally projecting head may act as a plane to give it increased 'lift'. Certainly it is a more agile shark than most. Some authorities declare that its head has evolved to make its detection of scent more accurately directional and its widely spaced eyes more effective, discounting the planing effect as secondary at best.

The marvellous design of the shark, the shape of its body and fins, arouses the envy of marine architects everywhere. They have calculated that the shark requires only one-sixth of the power per pound weight that a submarine of similar size would need to drive it through the water. It might seem that the fish's streamlining would be reduced by the roughness of its skin. Fishermen who have

handled catches of shark have suffered abrasions from their harsh sides. Cabinet makers in the past used its skin, which they called shagreen, to smooth wood. Unlike other fishes, sharks do not have flat, plate-like scales: they are covered with a surface of denticles. These are sometimes called placoid scales, and consist of pulp cavities that are covered with a layer of enamel, known as vitrodentine. The shape of the placoid scales and their grouping on the skin vary from species to species, but in effect, it may be said that a shark is actually covered in teeth. These small projections resist the flow of water as the fish moves through it, but it may be that they have the effect of increasing the fish's swimming efficiency.

Fishes experience two kinds of drag: that produced by the shape of the fish as it drives itself through the water, and another caused by

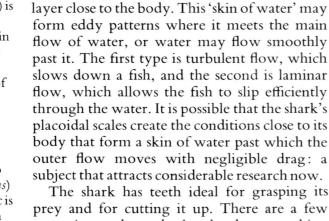

Left: The tiger shark (*Galeocerdo cuvieri*) is notorious for its attacks on bathers in shallow waters – it even enters river mouths in search of food.

Below: The mako shark (*Isurus paucus*) of the Indo-Pacific is a shark of the open ocean – it can leap right out of the water and has been known to attack small fishing boats.

the viscosity of the water itself where it lies in a layer close to the body. This 'skin of water' may form eddy patterns where it meets the main flow of water, or water may flow smoothly past it. The first type is turbulent flow, which slows down a fish, and the second is laminar flow, which allows the fish to slip efficiently through the water. It is possible that the shark's placoidal scales create the conditions close to its body that form a skin of water past which the outer flow moves with negligible drag: a subject that attracts considerable research now.

The shark has teeth ideal for grasping its prey and for cutting it up. There are a few exceptions where sharks also have crushing teeth: the Port Jackson shark (*Heterodontus phillippi*), a bottom feeder, has both cutting teeth and molars. Awl-shaped teeth are for grasping, and the triangular teeth are for shearing. They do not grow from sockets in the jaws, but are fairly loosely fixed in a sheath of connective tissue that is rich in collagen (a fibrous protein), and are secured in a kind of basal layer of the denticles. Behind the active rows of teeth there are five or six reserve rows. The shark, because its teeth are not tightly locked to its jaws, frequently loses a tooth in hunting. When this happens, a reserve moves forward to fill the gap. The teeth are also moulted quite often: a nurse shark (*Odontaspis taurus*) replaces its teeth every eight days. The great white and some other species have teeth with serrations which help them saw through chunks of their prey. Most sharks seize their food and, with a powerful shake of their heads and a twist of their bodies, tear a mouthful free to be swallowed whole.

The nearly supernatural ability of sharks to find suitable prey has produced many tall stories about their prowess. Their sensory equipment is truly wonderful. They sense their surroundings through their skin in ways that make it impossible for a man to imagine the shark's perception of its world. The lateral canal found in most fish is present in sharks, but does not involve the scales at all. The canals are filled with mucus and surface through pores in the skin, which is laced with many other canals. The shark probably detects subtle changes in pressure through them, helping it to control the laminar flow of water past its sides and picking up signals from other swimming creatures. A series of pore-like openings, the ampullae of Lorenzini, in front of its head, respond to the salinity of the water, changes in

pressure and temperature. They may also pick up the minute electrical impulses that radiate from the muscular effort of a struggling fish or a swimmer. In gaps between the placoid scales, small pits, rather like taste buds, remain a mystery: they may operate chemically and mechanically, noticing changes in pressure and the constitution of the water.

Like the wolf on land, the shark 'sees' much with its nose: it has been described as 'a swimming nose'. Two-thirds of its brain weight is accounted for by the olfactory areas. In fact the shark may 'taste' all over its body, a consideration that makes it especially hard for a man – a visual animal – to comprehend its sensations, and its less than humanly acute sight is something a shark might scarcely regret. Sharks were once believed to be unable to perceive colour, but since 1963, when marine biologists found cones in a shark's eyes, the possibility of sharks being attracted to swimmers wearing bright costumes has become a probability. The shark eye is less well supplied with ganglia than the human eye, but its capacity for adapting to the dark conditions it meets in deep water and at night is probably superior. Its hearing, too, has less range than human ears, but it is efficient over the cycles that really matter. A shark hears over a range of about 640 Hz – 10 Hz.

Man and the Shark
The exquisitely sensitive great white, cruising the oceans, rarely comes close to land: but when it does, its presence can close a resort's beaches in an afternoon. Few swimmers have encountered one, even a small one, and survived. In 1963, Rodney Fox, an Australian out for a day's skin diving near Adelaide was seized by the upper torso in the jaws of a great white.With remarkable presence of mind, he gouged the shark's eyes and escaped with multiple lacerations. His chances of survival would have been nil if he had met a group of sharks in the state of frenzy that affects them sometimes when feeding. Sharks in this state throw themselves at the source of food, tearing at each other and crashing into one another in an uncontrolled state. Sharks disembowelled by whalemen when they are in a feeding frenzy, have been seen to devour their own entrails before dying and being ripped to pieces by their companions. This is the behaviour that makes the shark the most terrible predator of all.

The Box Jellyfish
Terror of the ocean

An encounter with a box jellyfish, commonly known as sea wasps, in the tropical waters of northern Australia may well prove fatal, for these graceful marine coelenterates can inflict one of the most deadly of poisons in the animal world.

The Deadly Quartet

Box jellyfish belong to the Cubomedusae, a family of jellyfish distinguished by their rather box-shaped bell. The most dangerous members of the group belong to the sub-family Chirodropsidae. There are four main types; we can call them the deadly quartet or the four Cs: *Chironex, Chiropsalmus, Chirodropus* and *Chiropsoides*. Amongst these *Chironex fleckeri* ranks as the number one killer, accounting for over sixty fatalities a year along the Queensland coast.

An interesting feature of the group is the presence of a highly developed eye, one on each side of the bell. The eye consists of six eye spots made up of photosensitive cells but the largest of these has a lens, a very unusual feature among marine invertebrates and more typical of higher members of the Mollusca, such as the octopus.

A mature specimen of *Chironex* may weigh as much as 3 kilograms ($6\frac{1}{2}$ pounds) and have a bell 15–20 centimetres (6–8 inches) across. From each corner of the bell extend finger-like processes from which emerge up to sixteen extensile, almost transparent tentacles. At rest these measure from 90 centimetres to 9 metres (3–30 feet). Since there may be as many as sixty-four of these long invisible strands trailing beneath the surface of the water with their ends far removed from the main body of the jellyfish, it is easy enough to understand how a swimmer can be attacked without knowing what caused the injury.

These jellyfish are strong and graceful swimmers and are capable of propelling themselves at a steady 2 knots through the clear warm coastal waters. Swimming is effected by the alternate inhalation and exhalation of water through the 'mouth' situated on the ventral side of the body. During the summer months adult jellyfish swim at or near the surface, preferring the quiet, shallow waters of protected bays and estuaries with sandy bottoms – the ideal places for bathing! In fact several fatal accidents have occurred when people have been paddling in shallow water. The sudden, excruciating pain of an attack makes the victim jump up and move away, causing the adhering tentacles to be torn from the jellyfish and to wrap themselves around the victim causing further stings. The immediate effect, apart from the terrible pain, is the prompt formation on the afflicted parts of massive linear wheals with characteristic transverse bars. These wheals produce large red blisters and eventually form permanent scars on those who survive.

The Sting Mechanism

The organs that cause injury are the nematocysts, contained in batteries of single cells called cnidoblasts found on the surface and along the whole length of the tentacle. These cells, a characteristic feature of coelenterates, are also found in sea anemones, coral polyps and freshwater types like *Hydra*. Their main purpose is to stun small prey like fishes which are brought to the mouth region by the gradual retraction of the tentacles – but they also serve as extremely effective organs of defence. Each cnidoblast cell has projecting from the surface of the tentacle a trigger hair or cnidocil. When this touches an animate object it everts the nematocyst, discharging its fine tubule into the victim. Backward-curving butt spines at the base of each nematocyst help anchor the 'prey' so that it is quickly held and drawn into the range of more and more stinging cells. The poison is transmitted through numerous fine projections that are arranged spirally along the length of each nematocyst tubule. The combined action of thousands of discharging nematocysts can immobilize and kill the largest of animals. It is as though these jellyfish have been biologically programmed for 'overkill', having a supply of poison more than sufficient to kill small fishes, the normal prey.

Symptoms of Attack

When a bather has been attacked, if a less than lethal number of nematocysts has been discharged, damage may be lessened by carefully and quickly removing any remaining strands of tentacles. Rubbing the afflicted parts with wet sand or a cloth may do more harm than good by triggering off more nematocysts. It is believed that alcohol in any form, but preferably acidified with vinegar, if liberally applied is effective in dehydrating the gelatinous tentacles and deactivating the nematocysts. For the unfortunate victim, pain and scarring are just the first symptoms that may lead to death. Localized swelling, painful muscular spasms, respiratory distress, a rapid and weak pulse, and respiratory failure are the rapid consequences of an encounter with this jellyfish. Post-mortem examinations have shown that the lungs of the unfortunate victims are filled with large quantities of frothy mucus, with extreme congestion of the brain, kidney and abdominal organs. Most deaths occur from thirty seconds to fifteen minutes, but may take up to two hours. Those who have survived talk of the excruciating pain which causes them to writhe around on the ground, and they become completely irrational in their throes of agony.

The Medical Problem

The dramatic consequences of severe Cubomedusan sting have attracted the interest of many medical scientists and marine biologists. Research at the Commonwealth Serum Laboratories, Melbourne, is underway to produce an antivenin capable of limiting the severity of reaction of human beings to sublethal doses of poison from at least one species, the terrible *Chironex,* the species that presents the greatest hazard to swimmers in the warm waters of the Indo-Pacific. In the meantime the only way to swim safely in these waters is to wear a wetsuit or a suit made out of pantihose whose mesh is just fine enough to prevent penetration by the jellyfish's nematocysts. Apart from the species mentioned, the waters around Australia contain large numbers of other dangerous, if not fatal, stinging species including the 'giant blubbers' (*Cyanea spp.*) and 'bluebottles' (*Physalia spp.*). Jellyfish stings are a genuine medical problem in Australia where up to 3000 bathers have been treated at Sydney beaches on a single weekend. This alarming figure indicates the seriousness of such attacks.

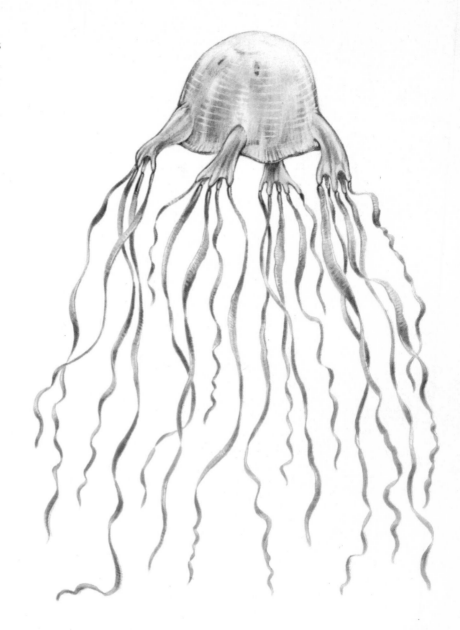

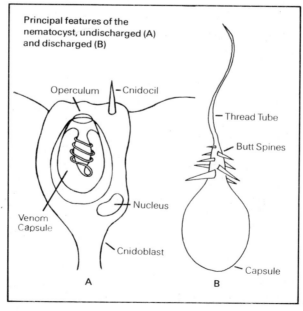

Principal features of the nematocyst, undischarged (A) and discharged (B)

Operculum — Cnidocil

— Thread Tube

— Butt Spines

Venom Capsule

— Nucleus

— Cnidoblast

— Capsule

A   B

Above: The box jellyfish or sea wasp (*Chironex fleckeri*) belongs to the order of jellyfish known as the Cubomedusae, distinguished by their cubic or box-like bells. From each of the four corners dangle up to sixteen stinging tentacles. These jellyfish are deadly, their venom being similar to that of a cobra but faster-acting. It can kill a human in less than one minute.

57

The Mosquito

Minute mass murderer

Far right: Emerging from their marshland breeding ground, a cloud of mosquitoes takes to the air. Only the females suck blood, the males feeding on the nectar of flowers.

Right: As she sucks blood to nourish her developing eggs, this *Anopheles* mosquito releases spores of the microorganism *Plasmodium* which multiply in the body of man and other vertebrate hosts causing malaria. The bite itself is painless, the irritation starting minutes after the insect has safely flown away.

Right: Malaria still causes over a million deaths each year although it has been eradicated from much of its former range.

The mosquito, a tiny two-winged fly, is the most dangerous insect in the world. This is because it carries the organism responsible for malaria, one of the most widespread and debilitating sicknesses of the tropics. At least fifty per cent of all human deaths since the Stone Age have been due to the bite of this insect. Even today some thirty years after an intense war against malaria, over one million people are killed by the disease every year and many millions more suffer from it. But the havoc and distress wrought by the mosquito in

transmitting malaria are only part of the story, for it carries other painful and deadly diseases including yellow fever, dengue or bone-break fever, filariasis which causes elephantiasis, as well as other viral diseases.

Although these diseases are associated with steaming jungles and tropical swamps mosquitoes themselves have a worldwide distribution and there are in Britain four species of this insect capable of transmitting malaria. Indeed 150 years ago malaria was quite common in Britain, especially in the low-lying marshy areas of Kent and East Anglia. In those days people would talk about an attack of 'ague', as they called it, just as people today speak of the severity and symptoms of each year's outbreak of influenza. The last notable outbreak of malaria in England occurred just after the First World War when soldiers carrying the disease returned home – those based in the area of the North Kent marshes acted as reservoirs of the disease which by means of the mosquito spread rapidly and infected half the population of the area.

Mosquitoes have existed on earth for at least fifty million years before man and have

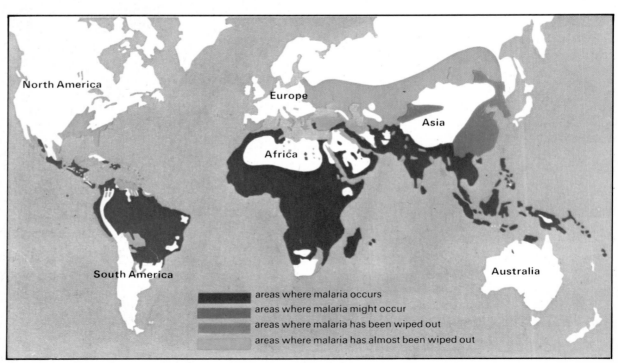

North America

Europe

Asia

Africa

South America

Australia

areas where malaria occurs
areas where malaria might occur
areas where malaria has been wiped out
areas where malaria has almost been wiped out

adapted to feed on a wide variety of animal hosts, from insects such as ants to reptiles and birds as well as a large number of mammalian hosts. Only the female mosquito bites and has mouthparts adapted for sucking blood. Rudyard Kipling's famous dictum 'The female of the species is more deadly than the male' is thus most apt in the case of mosquitoes. However, even many female mosquitoes can lay their first batch of eggs without the nourishment of a blood meal. Like the male throughout his life the female feeds first on the nectar of flowers.

Most mosquitoes lay their eggs on or in water in the form of small clumps which often resemble miniature rafts. The larvae feed and develop in water, undergoing three moults before they transform into the pupal stage which unlike that of a moth or butterfly is able to swim through the water. Mosquito larvae are commonly seen in water butts where if you disturb the surface they wriggle down deeper but soon come up again to breathe through tiny hair-like structures on their 'tail'. On average a mosquito spends three weeks in the water before emerging as the adult flying insect. After hatching mosquitoes gather into swarms that dance up and down in the air often under a tree or beside a large animal such as a horse or cow.

The patient observer will often be able to make out a swarm of somewhat larger mosquitoes dancing closer to the ground; these are the females. Suddenly a female will launch herself into the male pack – for a moment the males converge on her before a pair separates and sinks to the ground to mate. It is thought that the male detects the female by the different frequency of sound she produces when flying. The response to the female's high-pitched whine has caused some problems. In Canada the generators of a newly built power station were repeatedly blocked during the summer months with the bodies of millions of mosquitoes. After local entomologists were called in it was found to be due to the hum of the generators matching that of the female mosquito which acted as a super stimulus to male mosquitoes for miles around. The fault was rectified by altering very slightly the speed of the generators and the tone of sound they produced.

It is in northern Canada and throughout the Arctic Circle that the largest number of mosquitoes can be found at any one time.

These insects are a plague not only on the human inhabitants but also the native animals which serve as the main source of blood. Here on the tundra during the short summer the chief victims are nesting birds and the large, slow-moving musk ox which is bitten around the eyes and lips. In these northern parts the attack rate by mosquitoes on a man's forearm is 280 bites a minute which means that a totally unprotected man would lose half his blood in the space of a few hours. It is not uncommon to find the corpses of dogs and cattle killed by a particularly heavy mosquito attack. Despite the discomfort of these mosquitoes they are harmless compared to those that carry disease.

The Embassy of Death

For centuries people were stricken down and died of malaria without knowing the cause of the disease, although, as can be seen from its name meaning bad air, it was associated with the bad air seeping from swamps and marshlands. It was not until 1894 that the British scientist Sir Patrick Manson discovered that the mosquito was the vector of malaria.

In the process of sucking blood with her piercing mouthparts the female *Anopheles* mosquito releases malarial spores into the human bloodstream via her saliva. The actual causal agent is *Plasmodium,* a relative of the single-celled protozoan *Amoeba.* After release into the bloodstream *Plasmodium* spores enter the red blood cells where they multiply and burst back into the bloodstream before entering other hitherto uninfected cells. As the spores break out and rupture they release toxins which cause the fever associated with the disease. When a second 'clean' mosquito bites an infected person she takes up with the blood some of these spores where they undergo further development in the mosquito's stomach and later in her salivary glands. The World Health Organization, the main body responsible for combating malaria, has described the symptoms of malaria:

> First there is shivering; the chattering of teeth, convulsive fits, a skin icy to the touch and a temperature of 104 or higher. After the shivering comes the burning dry heat that drives you almost insane, the insatiable thirst, the booming of the head, the delirium and, worst of all, the hot prickly fire on the skin that is like an excursion into hell.

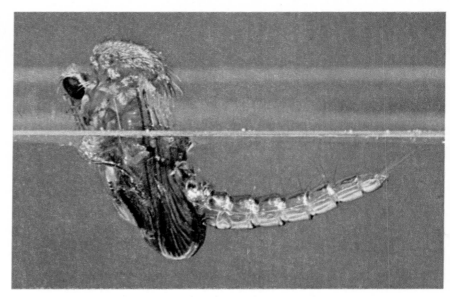

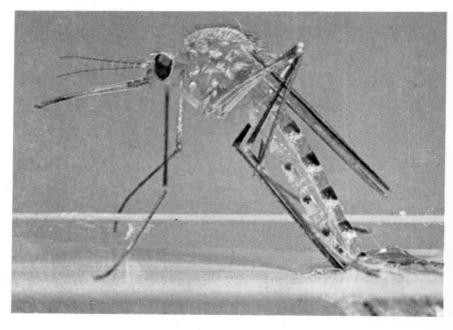

Often the disease develops into black-water fever, named from the dark colour of the urine due to the number of damaged red blood cells.

Malaria, named by the ancient Indians as the 'King of Diseases', has wiped out whole civilizations and even the fall of both the Greek and Roman Empires has been attributed to it. It was not until the sixteenth century when the Spanish brought back to Europe the drug quinine (obtained from the bark of the cinchona tree) that some source of relief was found to lessen the effects of malaria. But it was many years later before anything could be done effectively to combat the disease. After the discovery of DDT in the 1940s the World Health Organization launched a massive anti-malarial campaign attacking both the adult mosquito and its breeding grounds as well as aiming to save the lives of those people already infected. Since the fifties there has been notable success in India, the Caribbean and Venezuela, parts of South-east Asia and in the vestigial pockets of the disease in southern Europe – but much of tropical Africa, the original 'white man's grave' remains rife with it. Although most of Europe today is immune, in 1958 10,000 people died of malaria in Greece. Perhaps the worst epidemic this century occurred in Sri Lanka (Ceylon) when in 1934, after the disease had almost died out, it flared up in massive proportions during the climax of a drought when all the residual pools of water were heavily infested with the larvae of the *Anopheles* mosquito. Over 500 people a day died and it took thirty years to control the disease. Unfortunately the campaign that eliminated the disease so successfully ran out of funds in the 1960s and at any time it could erupt and spread again among the population. It should be added that government agencies, when helping to advertise their exotic countries as the place for a unique holiday, often fail to mention malaria and the many other diseases one can contract within a stone's throw of those golden beaches!

Whereas several species of *Anopheles* mosquito are potential carriers of malaria only one species, *Aedes aegyptii*, is the vector of the deadly yellow fever. Here the agent responsible is not a protozoan but a virus. Africa is the centre of distribution of this disease and it is thought that it was spread across the Atlantic to South America by the slave trade. It causes stomach haemorrhage and liver damage pro-ducing jaundice and the typical yellow colour of the patient. While yellow fever is associated with the tropics it has during the last 200 years decimated the populations of many North American cities notably New Orleans, Phila-delphia and Boston. The same species of mosquito that transmits yellow fever also carries dengue or bone-break fever which though seldom fatal causes excruciating pain in the back, neck and limb joints.

Other Diseases

To add to the horrors wrought by the mosquito is filariasis, a disease due to the infestation of the human body by tiny nema-tode worms. During the day these worms live in fluid-filled spaces in the lungs but at night they migrate to the bloodstream where they are picked up by the biting mosquito. The condition known as elephantiasis where the lower parts of the body swell up and assume gigantic proportions is due to a heavy infesta-tion of worms blocking a lymph duct. Cases are known of elephantiasis where the scrotum has weighed around 37 kilograms (81 pounds) and has had to be carried on a wheeled cart in front of the patient.

As well as the three well-known diseases mentioned so far there are innumerable viral infections known under the general heading of encephalomyelitis which are transmitted by mosquitoes. In a number of cases man does not appear to be the primary host but as popula-tions of mosquitoes build up in numbers man becomes a secondary host and suffers sudden outbreaks of what at first appear to be mystery diseases. Many of the fevers contracted by holidaymakers in southern Europe have also been attributed to mosquitoes. Migratory birds are one source of infection and biologists in France and Spain when ringing these birds to trace their winter and summer movements are often stricken with a fever which probably originated in the African bush.

The faint whining of a mosquito at night is an infuriating and frightening sound, for the least damage this tiny insect can inflict is to leave its victim with an irritating and unsightly bite, and despite nets, window meshes and the latest insect repellent it persists in its vocation to extract the drops of blood necessary for the continuation of its species. Even if man wiped himself out, the mosquito would survive . . . as it has done for many millions of years.

The Tiger
Lord of predators

The tiger (*Panthera tigris*) is the largest and most massively built member of the cat family. It is also the most adaptable in the large range of habitats it occupies. Tigers live in the snow-covered forests of Siberia (where temperatures can drop to −35°C), on high passes in the Himalayas (to above 3000 metres, or 10,000 feet), in swampy areas with dense cover and in the dark, humid jungles of Indonesia. Of the eight recognized sub-species the Bengal or Indian tiger has the most fearsome reputation as a man-killer. 'At a conservative estimate tigers have consumed well over half a million Indians in the past four centuries. . . . Entire districts have been depopulated and villages abandoned, in some cases for years because of the work of man-eaters.'

The Hunter Hunted

The most vivid and exciting accounts of ferocious tigers come from the writings of Colonel Jim Corbett, otherwise known as 'Tiger Jim'. In the early years of this century Corbett killed singlehandedly half-a-dozen man-eating tigers, not to mention the numerous rogue leopards who attacked at night whereas the tigers killed predominantly during the day.

Of all tigers, the Champwat tigress is the most famous, for this one animal was said to have killed 438 people before Corbett tracked her down and shot her in 1911. By the time he got onto her trail she had already taken 200 lives in Nepal before she was driven across the border into Kumaon Province in northern India. It was here in the foothills of the Himalayas that she terrorized villagers, attacking at one spot and then again, swiftly and silently, a few miles away. The local people had no firearms and were helpless against her attacks until Corbett came along. Most of the victims were young women who spent some time away from the village cutting fodder for the cattle. The tigress appeared not to be intimidated by a group of such women and, choosing a victim, she would strike and carry her away from within a few feet of her companions. Other casualties occurred among men who strayed away from the village alone in order to gather or chop wood for fuel.

Although Corbett admits he was paralysed with fear he describes the chase with a coolness and precision almost worthy of Sherlock Holmes. At the same time the impression of frantic activity is created, as he scrambles up rocky screes, through beds of nettles and down steep-sided gorges with the agility of a mountain goat. He does, however, display an understandable pathos when describing the state of the unfortunate victims' families. Referring to the site where a young woman had been killed by the Champwat tigress, he writes: 'The spot where the girl had been killed was marked by a pool of blood and near it in vivid contrast to the crimson pool was a broken necklace of brightly-coloured blue beads.' Hot on her trail he pursues the tigress to the banks of a stream where, forced to move on, she leaves further gory evidence.

> Splinters of bone were scattered round the deep pug marks into which the discoloured water was now seeping and at the edge of the pool was an object that puzzled me as I came down the watercourse, and which I now found was part of a human leg. In all the subsequent years I have hunted man-eaters I have not seen anything more pitiful as that comely leg – bitten off a little below the knee as clean as though severed by the stroke of an axe – out of which the warm blood was trickling.

G. P. Sanderson, another expert of Indian wildlife, gives an equally harrowing description of a victim of a tiger attack which suggests the immense strength of the beast as it strikes out at its prey.

> The next death was of a horrible description . . . several villagers were grazing their cattle in a swampy hollow when the tigress pounced upon one man. She missed her aim at his throat, seized the shoulder, and then, either in jerking him, or by a blow, threw him onto a thicket several

Hunting and Feeding

feet from the ground. Here the wounded wretch was caught by thorny creepers; whilst the tigress, as generally happens when a *contretemps* takes place, relinquished her attack. Next morning the lacerated wretch was found. In his mangled state he had been unable to release himself; he was moaning and hanging almost head downwards amongst the creepers; and died soon after he was taken down.

Corbett emphatically believed that all man-eating tigers are wounded or sick individuals; for example he found that both the upper and lower canine teeth of the Champwat tigress were broken due to a gunshot wound. Another man-eater had suffered in an attack against a large Indian porcupine and had lost an eye as well as having numerous lacerations and quills embedded in her paws.

Tigers have large appetites and require on average 6 tonnes of meat a year to sustain themselves. Schaller, the eminent mammalogist, who has done much work on the tiger and its prey, estimated that a man-eating tiger would have to kill sixty people a year – if resorting to an exclusively human diet. The Champwat tigress appeared to have agreed with his figures having taken on average fifty-six people each year while rampant.

Depending where they live tigers take a wide range of food. In India the natural prey is the elegant chital deer, blackbuck, hog deer and wild boar but in the last fifty years many of India's larger wild animals have become quite rare and since then the tiger has turned its attention towards cattle and domestic buffalo. When stalking, the tiger behaves just like a domestic cat as it creeps nearer and nearer to its prey, half crouching and quivering with ex-

Above: A tiger's warning growl exposes its formidable canine teeth – the tiger is the largest and most powerful land predator.

Overleaf: A group of Indian tigers in a Malaysian Wildlife Park. There are approximately eighty of these animals held in captivity and up to 800 of the larger Siberian race.

Above: The size difference between the sexes is noticeable in this picture of a pair of Siberian tigers. Males of this race are massive animals – the largest of all the big cats, growing to well over 4 metres (13 feet) in length.

citement, its striped body being beautifully camouflaged against the grasses and the dappled sunlight filtering through the bushes and dense vegetation. As soon as it gets within range (10–25 metres or about 30–80 feet) it runs and springs, often landing on the back of the prey in order to bring it down. The fangs bite deep into the neck while the claws of the forepaws dig into the face and the muzzle. If the nerve cord is not severed, the tiger quickly changes its grip and bites at the throat. Cattle are relatively easy quarry but buffaloes will often bunch together to defend themselves or to ward off a tiger after one of the herd has been killed. Wild boar, especially adult males, can also defend themselves by lashing out with their tusks. Evidently tigers are not the most efficient of hunters but their persistence brings results.

Once the prey is killed it is dragged or carried in the mouth into cover and often near water. The amazing brute strength of the tiger is shown by the fact that the carcase of a domestic buffalo has been dragged over 76 metres (250 feet) without a pause, and a cow seized in a native compound has been carried over a 2-metre (6½-foot) wall with apparently little effort. Human victims have been reported to be carried in the mouth as a dog would carry a large rat. An average first sitting for a tiger consists of about 32 kilo-

grams (60 pounds) of meat, after which it usually heads towards a stream or pool where it drinks copiously before laying out near the kill. To prevent attacks by scavenging jackals and vultures a tiger often covers the remains over with earth and vegetation.

There are some seemingly tall stories of tigers bringing down elephants but from the many eye-witness accounts these do appear to be true. Often two tigers are involved, generally a tigress and a fully grown cub. Apparently the tigers surprise the elephant by leaping onto its back and lashing out with their claws at the elephant's eyes while the fangs are buried in the poor animal's neck and throat. On the other hand tigers have been reported to have been mangled and trodden into a bloody pulp by female elephants enraged after the loss of their calf. A half-tonne armour-plated Indian rhinoceros has even been reported to have succumbed to the ferocity of a tiger's attack. In the north the Siberian tiger is known to dig out brown bears from their dens and kill both mother and cubs. Wolves are also reported to have been eaten. Apart from man the tiger's only real enemy is the hunting dog. Packs of these animals will trail and succeed in killing a tiger although it may slay several members of the pack before it is finally overcome by sheer numbers.

Conservation

Today the tiger is classified as endangered. By the 1930s the world population of tigers had dwindled from over 100,000 to less than 7000. Hunting the tiger for sport, the loss of its natural prey and wild habitats have all contributed to its decline in numbers. At the time of the British Raj tiger hunting was a favourite pastime of army officers but the numbers bagged by the British were negligible compared with those of the Indian maharajahs. For example the Maharajah of Udaipur was said to have shot at least 1000 tigers. It should be added that this sport was not as one-sided as it would appear, for a tiger would often leap at an approaching elephant and snatch the howdah from its back before the hunting party had a chance to retaliate.

In 1972 the World Wildlife Fund launched a massive fund-raising campaign called 'Operation Tiger' to rescue the tiger from what seemed at the time to be inevitable extinction. This successful operation raised over £800,000 within eighteen months, much of the money being contributed by schoolchildren from Switzerland, Great Britain and the Netherlands. The money has been spent strengthening existing reserves, creating new ones, conducting population studies and monitoring the behaviour and hunting requirements of the tiger. The work has concentrated mainly on India and the neighbouring states of Nepal, Bhutan and Bangladesh where together there are over a dozen tiger reserves. Already these animals have increased in numbers and if efficiently maintained their survival is assured. Protection measures came too late for the Balinese race which became extinct in the late seventies. The Javan race appears to be following the same path, having been reduced to only a couple of individuals, and it is doubtful if the Caspian race still survives.

While man protects the tiger, the tiger continues its man-eating habits. The most dangerous area today for man-eaters is the vast labyrinth of mangrove swamps and tidal waterways known as the Sundarbans in West Bengal, which forms the delta of the two great Indian rivers, the Ganges and the Brahmaputra. Here tigers have killed 300 people in the last fifteen years. The victims are most often solitary individuals who paddle across the Bay of Bengal from the mainland to collect wild honeycomb and wood from the dense thickets of vegetation on the numerous islands. Tigers here attack both on land and in the water, as they are powerful swimmers and can easily capsize a small boat. The last tiger to make headlines was shot at Ramiri Udaygiri in India by an American woman hunter. It was claimed to have killed and eaten 500 people over a six-year period but this is thought to be a gross exaggeration. There must, however, be some truth in the story, and it is certain that many of these deaths are attributable to a lack of natural prey. In the end, therefore, it is man who suffers for his continued though indirect persecution of the tiger.

Below: One of the favoured habitats of the Indian tiger is the great wetland region of the Sundarbans in West Bengal where most man-killing tigers are found.

The Vampire Bat

Blood thief of the night

Below: A pair of common vampires (*Desmodus rotundus*) are disturbed during their daytime roost. One of the bats exposes its fine chisel-like teeth, used to painlessly cut through the skin of its victim.

Blood is a substance that most of us do not care to contemplate too closely, especially if it is our own, and a creature that lives by stealthily feasting on it is hard for us to love. Nevertheless, the horror-filled legends that have grown up about the vampire's choice of diet are far from fair to this small bat. In the nineteenth century an Irish novelist, Bram Stoker, who probably cut his teeth on tales of Rudy Rawbones and other discouraging Irish bogeymen, created the character of Dracula, a 'human vampire' of supernatural powers. He lent credence to his creation with stories of giant vampire bats that rose from the ocean to drink the blood of helpless seamen, draining these poor fellows dry.

Legends of blood-drinking ghouls have been common all round the world for many centuries. The Slavs, especially, have a rich tradition in such beliefs, perhaps encouraged by their custom of two-stage interment of their dead: the decomposition of the first burial being hastened by burning fires on top of the grave and pouring water on it, then the disinterment to burn the bones later. When for some freakish reason the corpse was found to have resisted corruption, the presence of a vampire was diagnosed, and the Church was called in to exorcize the demon.

The true vampires, found in Central and South America and the island of Trinidad, are far less dreadful. Being no longer than 9 centimetres ($3\frac{1}{2}$ inches) from nose to rump, the only animals they might be able to drain of their blood at one visit would be birds or small rodents. In fact, this does happen occasionally, but the vampire's prey is generally much larger and lives to provide many more meals for the midnight diner.

For many years people accounted for the strange way in which the vampire's victims were apparently complacent about the bat's assaults by suggesting that the bat wafted a soporific scent towards the nose of its hosts, lulling them into a deep sleep. It is true that bats have scent glands, but anyone smelling their products is unlikely to be spellbound by their perfume, which is rather like a whiff of ammonia. The simple truth is surprising.

Vampires leave their caves or hollow trees after dark and before midnight. They fly on a wingspan of about 38 centimetres (15 inches), hugging the ground at a height of around 1 metre (40 inches). When they see a suitable prey, usually a horse, goat, pig or other domestic livestock, they land and walk silently up to it. They climb or flap their way noiselessly up the animal to a part that is unprotected by thick hide or dense hair before settling to feed. Their approach is so wonderfully stealthy that even experienced human watchers have been staggered to find that after a night's observation

Right: Vampires leave their roosts before midnight, flying low over the ground in search of warm-blooded prey.

they had failed to see the bat, although it had left the clear evidence of its visit. Its rather patchy, dull grey-brown fur gives it excellent camouflage, and the soft pads on its limbs make its touch the merest caress.

Once in position on its dozing, warm-blooded host, the vampire performs its small miracle of exactness in chiselling away a narrow strip of skin. The teeth it uses for this operation are the sharp incisors in its upper jaw. The vampire then extends its long tongue and laps up the blood. Its saliva carries an anti-coagulant substance that helps to keep the blood flowing. The vampire takes a more active part to encourage the supply of blood by curling the sides of its tongue downwards to form a tube. Working this 'tube' back and forth gently in the wound, a partial vacuum is formed and the blood is drawn into the bat's throat as if it were sucking it up a straw. If the supply seems to fail, it stimulates the cut with

Above: Once a sleeping animal such as a cow, goat or man is detected, the vampire drops down close by and stealthily creeps up on its prey searching for a suitable hairless part of the body such as the toe, muzzle or ear.

the tip of its tongue until the flow is restored. The bat's mouth is beautifully evolved for this method of eating: the lower jaw is deeply grooved for easy action of teeth and tongue.

When the vampire has finished its meal, it flops to the ground. Frequently, it is so gorged with blood that it finds great difficulty in taking to the air again. On many occasions, vampires have been seen lying a few metres from their prey somnolently digesting their food before returning to their roosting places.

The vampire bat is a close relative of the leaf-nosed bats, although its own nose is not very developed in this way, looking more like a pug nose than the extended leaf of, say, the red fruit bat (*Stenoderma rufum*). There are three

genera of vampires, each with one species. The common vampire (*Desmodus rotundus*) is, as its name suggests, the best observed. It colonizes caves, sometimes in association with other bats.

Man and the Vampire

The bats hunt their food in a fairly small area so the presence of a large colony can anger local stockbreeders greatly. An individual vampire rarely takes more than 28 millilitres (1 fluid ounce) of blood from its victim at a time, but if one multiplies that figure by the feeds that a bat will need for a year and adds perhaps a thousand other bats' takings, the figure rises to nearly 25,800 litres (5,675 gallons), which makes a farmer reach for his cannisters of cyanide gas. The wounds inflicted by the bat are not serious in themselves. The look much worse than they are. In the morning, however, an animal may still be bleeding, even for as long as eight hours after the incision has been made, so the loss of blood may be quite spectacular. More seriously, the bat carries the infections of murrina disease and rabies. In countries where these are endemic, the bat will pick them up and transmit them before dying of them itself.

It would be sad to lose this interesting and highly specialized creature, but efforts to exterminate it have been made in parts of Mexico, where cattle farming has suffered losses attributed to the bat's habits. Its numbers elsewhere seem to stay healthy.

Poisonous Amphibians

Deadly midgets on the forest floor

Below: The bold coloration of the European fire salamander (*Salamandra salamandra*) advertises that it is poisonous. Its skin glands contain a secretion which makes it distasteful and which is especially irritating to the mucus membranes. However, this amphibian is frequently preyed upon by the common grass snake which appears immune to the noxious secretion.

All amphibians have poison glands but the types and strengths of poison vary greatly between species. Even the common toad when provoked can secrete an extremely irritating poison from glands situated behind its eyes. A dog will retch and rub at its eyes after an abortive attempt at biting a toad and people who have handled toads roughly often break out in a rash and are overcome with a distinct feeling of nausea. Although not an absolute rule, many of the more poisonous amphibians advertise their distastefulness with a striking range of warning colour patterns. The black and yellow European salamander is a common example. This creature is quite harmless when handled but if attacked by a predator it secretes a strong poison capable of killing a small mammal such as a shrew or weasel.

The most deadly of all amphibians, however, are the frogs of the family Dendrobatidae, confined to the rainforests of Central and South America where they live mainly on or near the forest floor. These tiny frogs, which seldom grow over 2.5 centimetres (1 inch) long, exhibit a staggering range of colour and body patterns – from maroon with blue spots to bright pink and yellow. Until recently the most dangerous of all was thought to be the black and orange *Phyllobates bicolor*. However, in 1973, a tiny golden-coloured frog of the same genus was discovered in the lowland rainforest of Colombia. This frog was given the specific name '*terribilis*' on account of the extraordinary toxicity of its skin secretions. Although each frog contains only 1.9 milligrams (0.00007 ounces) of poison it is at least twenty times more potent than that of *P. bicolor*. The poison of one frog is capable of killing over 20,000 men! The active constituents of the venom are batrachotoxin and homobatrachotoxin, among the most potent of all naturally occurring poisons. They cause a chemical imbalance of salts at the nerve and muscle cell membranes, producing general

muscle spasm and heart failure. Other symptoms shown by laboratory animals include violent convulsions, salivation, and extreme difficulty in breathing. A mere trace of this poison transferred to the skin can cause a severe burning sensation.

A Game of Darts

Phyllobates bicolor, the recently discovered *P. terribilis* and the even smaller *P. aurotaenia* are all used by the Chocó Indians of Colombia to smear the tips of their poison darts, which are then inserted into blowpipes and used to kill large animals such as bear, jaguar, pig and deer. These darts also played a part in fighting the invading Spanish forces in the sixteenth century, who must have been terrified by the constant threat of swift, silent and cruel death lurking behind the trees. So strong is the venom that a man shot with such a dart can only stagger a few hundred metres at most before dropping dead.

The first written account of the use of poisoned darts came from Captain Charles Stuart Cochrane who travelled through the Colombian Cordilleras when on leave from the British navy in 1824. Although plagued by fever he kept a detailed journal which was published in two volumes. This excerpt describes how poison is extracted from the frog *P. bicolor:*

> Those who use the poison catch the frogs in the woods, and confine them in a hollow cane, where they regularly feed them until they want the poison, when they take one of the unfortunate reptiles, and pass a pointed piece of wood down his throat, and out at one of his legs. This torture makes the frog perspire very much, especially on the back, which becomes covered with white froth: this is the most powerful poison that he yields, and in this they dip or roll the points of their arrows, which will preserve their destructive power for a year. Afterwards below the white substance, appears a yellow oil, which is carefully scraped off, and retains its deadly influence for four or six months, according to the goodness of the frog. By this means, from one frog sufficient poison is obtained for about fifty arrows.

The technique of staking the frog down appears to be the most common amongst the various tribes of Indians, and for this they use a special bamboo peg called a *siura kida*. The pegging out induces stress which causes the frog to release its poison. When using *P. terribilis* there is no need to induce stress because the body literally oozes with poison. All that is required in this case is for the frog to be immobilized on the ground and the darts lightly scraped along the skin of the back.

Gains and Losses

Apart from their toxicity, frogs of the family Dendrobatidae show interesting reproductive behaviour. The male of each species has a distinct calling note which ranges from a long melodious trill to a short staccato buzz. Unlike the common frog, in which the male grips the female in a tight embrace (known as amplexus), fertilization is effected by the couple lying back to back with only their cloacas touching. The eggs are laid in small batches often in an upturned leaf which has collected rainwater, and the eggs and young tadpoles are guarded by the male. When the tadpoles reach a certain size, they slither up the male's back and attach themselves to a patch of sticky mucus secreted from glands in his skin. The tadpoles are then carried to a more permanent area of water such as a pond or ditch.

Below: *Dendrobates leucomelas,* one of the many strikingly coloured poisonous frogs of South America.

The Nile Crocodile

Creature of nightmare

Once a child sees a picture of a crocodile in a book he or she seldom forgets it and repeatedly asks how big do they grow, what do they eat and where do they live? One of the most common questions is whether they are found in England! Children seem to have an innate fear of crocodiles which is, after all, quite justified. After letting their imagination run riot it must therefore come as a great disappointment to see crocodiles in a zoo dozing in a sluggish torpor and barely moving an eyelid. These artificial conditions certainly do not do justice to such powerful predators who, when hunting in the wild, strike out with lightning speed.

Most of the harrowing stories of man-eating crocodiles are probably quite authentic, but only three of the twenty-one species of crocodile can truly be classed as man-eaters. These are the Nile crocodile (*Crocodylus niloticus*) which inhabits Africa (but was once found as far north as Caesarea in Israel), the mugger (*C. palustris*) which lives in India (and feed on the corpses thrown into the rivers for burial), and the estuarine or saltwater crocodile (*C. porosus*) which ranges from India east to the Solomon Islands and the swamps of northern Australia. The estuarine crocodile is not only the largest crocodilian, it is the largest reptile in the world, growing up to 9 metres (30 feet) and weighing over 2 tonnes. Although there have probably been more single incidents of Nile crocodiles attacking people, the estuarine crocodile holds the disgusting record of most mass kills. One of the heaviest tolls of casualties occurred during the Second World War when Japanese troops were forced by the British onto a marshy island just off the Burmese mainland. Of the thousand men trapped on the island only twenty survived a night attack by crocodiles, and the survivors told of the horror they felt as one by one their comrades were dragged into the inky blackness. Even the British troops moored some way off heard the victims' terrible screams. More recently, in 1975 forty-two people were eaten by crocodiles when their 'pleasure' craft sank in a crocodile-infested river in Indonesia.

The Nile crocodile is the one which has been written about most and has become something of a legend. Despite a reduction in numbers caused by hunters after their skins, these reptiles still account for at least a thousand human deaths a year in Africa. They are also responsible for decimating herds of cattle which are either snatched at the water's edge or pulled under as they cross a river. However, it takes many years for a crocodile to grow to a sufficient size to take a man or any other large mammal.

Survival of the Fittest

The Nile crocodile starts life as a small replica of the adult, about 30 centimetres (1 foot) long. The mother, hearing the high-pitched 'cheeping' of the young within the eggs aids their exit from the nest by scraping away the compacted earth in which she had buried them. After hatching she carries them with inordinate care in her mouth to the water. After following the mother around for a few days like small ducklings the young soon disperse for cover. Only one per cent of these crocodiles survive their first year; the rest fall prey to a host of predators including monitor lizards, otters, herons and eagles. The young crocodiles' first food consists of insects like large water bugs, dragonfly larvae and soft-bodied frogs and toads. When little over 1 metre (3 feet 3 inches) long, the young are expert swimmers and their diet consists mainly of fish including catfish, lungfish and numerous kinds of tilapia. As they grow they take larger prey such as ducks, which are grabbed by the feet while swimming, or other water birds surprised at the water's edge. It is only when they are over 2 metres (6 feet 6 inches) that they are powerful enough to take larger mammals like otters and young antelope.

The bulky adults are slow swimmers and have to rely more on stealth and camouflage for their food. These massive adult crocodiles become experienced in the habits of their prey and seldom fail to make a kill when hungry. As dusk falls crocodiles become active and move in towards the banks where they wait among the shallows for animals coming down to drink. When an animal such as an antelope comes within striking range the crocodile shoots forward like a torpedo, sinks its large teeth into the animal's muzzle or forefeet and quickly swivels round to knock the prey off-balance. If successful it drags it quickly down into the water where it soon drowns. It was thought that crocodiles stored their food after killing in order to let the flesh putrefy and make it more manageable but, in fact, a crocodile is quite capable of eating fresh meat. When feeding it latches its jaws onto the soft under-parts of its prey and, spinning round full circle, it wrenches off large chunks of meat which are swallowed with jerky movements of the head.

The most common mammalian prey consists of marsh antelopes – waterbuck, sitatunga and lechwe, but all mammals are taken, even water buffalo, rhinoceros and giraffe. The remains of hippopotami have also been found in crocodiles' stomachs although, in turn, many crocodiles have been found mutilated by hippos' massive teeth. The only animal capable of holding its own against a big crocodile is an elephant and there have been several cases of a crocodile attacking an elephant calf and being battered and trampled to death by the irate mother.

Most human casualties in Africa consist of children and women who spend a lot of time gathering water and washing clothes at the water's edge. Some tribes build protective stockades but most native people accept the crocodiles' presence as part of the landscape and they tend to become familiar with the feeding habits of crocodiles on their particular stretch of water.

One of the most frightening aspects of crocodile behaviour is their habit of chasing their prey out of the water all the time lashing out vigorously with their tails to knock the animal off-balance. There was one tragic case in which a young African in a canoe was pursued by a crocodile but managed to land on an island in the middle of the river. However, the crocodile quickly scrambled after him up the bank and dragged him screaming back into the water.

Below: The resemblance of the Nile crocodile (*Crocodylus niloticus*) to a floating log allows it to approach its prey without detection.

Overleaf: The massive gape and formidable armament of teeth make the Nile crocodile one of the most powerful of freshwater predators. This reptile is a true man-eater and will tackle prey as large as giraffe and rhinoceros.

The Scorpion
Pincers of fear

Armed with powerful pincers and a sting at the end of the tail these shuffling, skulking arachnids inspire fear in all who encounter them. But despite their reputation most scorpions are small and relatively harmless, their stings having as much effect as that of a bee or a wasp. The largest species live in India and tropical Africa and measure up to 30 centimetres (1 foot) in length from the head to the tip of their tail. Although these large specimens can administer a massive dose of venom their sting is seldom fatal. But a number of scorpions are dangerous, even deadly. The most dangerous of all in that it is responsible for most deaths is the North African species *Androctonus australis* whose venom is as potent as that of a cobra. This scorpion can kill a dog within a matter of minutes and a man in the short space of four hours. Fortunately an antitoxin has been developed for this species.

The type of poison administered by *Androctonus* is termed neurotoxic as it primarily affects the nervous system but it also has a haemolytic action in destroying the red blood cells. After the initial sharp jabbing pain at the site of the sting the infected arm or leg becomes numb, the muscles start twitching, there is difficulty in breathing followed by slurred speech and loss of bladder and bowel control. The victim perspires profusely and the skin becomes cold and clammy to the touch. In fatal cases victims have been known to froth at the mouth and before death the limbs turn blue due to the destruction of red blood cells.

In Mexico and America's south-west there are two other deadly species of scorpion of the genus *Centruroides* which between them kill on average a thousand people a year, the main casualties being children who run around barefoot. Even in Europe one can encounter scorpions – one of the hazards of a camping holiday in Spain or the South of France is the small yellowish *Buthus occitanus* which often finds shelter under the groundsheets. Its sting, although not deadly, can be extremely painful. The other common European scorpion is the small black *Euscorpius flavicaudis* which tends to live in damp places.

Below: The sting of the fat-tailed scorpion (*Androctonus australis*) is extremely painful and can kill a man within a couple of hours. This species is widely distributed throughout North Africa.

Love with a Sting in the Tail

Despite their formidable weaponry scorpions only sting when provoked and they are not aggressive by nature. Most species are nocturnal, emerging at dusk from under the shelter of a stone or a shallow burrow which they excavate themselves with their walking legs. Among themselves scorpions could be said to have a loving nature, up to a point. Prior to mating or fertilization of the eggs an elaborate courtship ritual is performed which involves the male grabbing the female with his pedipalps (pincers), whirling her round in a dance and walking her up and down in a kind of nuptial promenade. In an integral part of this courtship dance the partners face each other, press their abdomens flat on the ground and extend their tails upwards repeatedly intertwining them with that of their partner.

After the whole performance, which may last several hours, the male deposits a packet of sperm (known as a spermatophore) on the ground and drags the female backwards until her sexual opening takes up the sperm packet. After the eggs are fertilized, however, love flies out of the door and the male is often attacked and eaten by the female. The fertilized eggs develop inside the mother and once the soft-skinned young hatch out they immediately clamber onto their mother's back where they remain for a few days. The French naturalist, J. H. Fabre, who observed scorpions over several years in a purpose-built enclosure, describes in detail the highly developed maternal instinct of the female scorpion:

So here we have the young nicely wiped, clean and free. They are white, their length from the forehead to the tip of the tail measures 9mm ($\frac{1}{3}$ inch). As the remnants of the egg are discarded, the young climb, first one and then the other, onto the mother's back, hoisting themselves, without excessive haste along the claws, which the mother scorpion keeps flat on the ground, in order to facilitate the ascent.

The natural diet of scorpions are insects like beetles and cockroaches which are caught with the pincer-like movements of the pedipalps. The prey is dismembered by the paired mouth appendages, the chelicerae, and the juices and soft tissues are drawn into the mouth by a pumping action of the pharynx. The poison sting is only brought into play when the prey is large and puts up a struggle.

Not-So-Poor Relations

Wind-scorpions, camel spiders and solfugids are just a few of the names given to a strange group of animals related to the true scorpions. These arachnids have in proportion to their size the largest jaws in the animal kingdom, capable of crushing and dismembering any hard-shelled insect. They will also tackle prey much larger than themselves like gecko lizards and small rodents. Once the prey is secured the two pair of jaws or chelicerae work together like jagged-blade secateurs to mince it up in a matter of seconds. An adult wind-scorpion has been seen to eat one hundred large flies in a day – these animals are notorious for gorging themselves until their abdomens become so distended they are unable to pursue further quarry. Capturing one of these creatures can be difficult, as it will back into a corner, raise its head and thrash its jaws furiously. Although they lack poison fangs their well-developed jaws can inflict a nasty wound.

For all their ferocity camel spiders are quite small and never grow over 15 centimetres (6 inches) long – but they appear larger because their bodies are usually clothed with numerous stiff white hairs. In South Africa, the centre of distribution, they are feared not only for their bite but the lightning speed at which they can run. They can be seen in large numbers in the Namib Desert of south-west Africa dashing across the dunes like small balls of white thistledown.

Left: The jaws of the wind-scorpion or camel spider are, in relation to its size, the largest and most formidable of all animals.

POISONOUS ANIMALS AND THEIR VENOMS

Name	Method of venom delivery	Type of venom
Box jellyfish *Chironex fleckeri*	This jellyfish has four bunches of tentacles covered with special cells which when touched discharge hollow stinging threads, loaded with the venom.	Not known, but includes a histamine or histamine producer, a pain-producer and substance causing paralysis of the heart and especially the respiratory organs.
Geographic Cone *Conus geographus*	One of a group of beautiful shelled molluscs which have harpoon-like hollow teeth stored inside a sac within the mouth, which is fired at the prey.	Unknown, but similar in action to curare, acting directly on the skeletal muscle and causing paralysis.
Blue-ringed octopus *Hapalochlaena maculosa*	Modified 'salivary' glands producing venom open into the mouth of this octopus, and the sharp horny parrot-like jaws introduce the poison into the prey.	A protein which has been named 'cephalotoxin' has been extracted from several kinds of octopus and shown to cause paralysis in crabs and rabbits.
Arizona Scorpion *Centuroides sculpturatus*	Scorpion venom is made in paired glands near the end of the tail. An attacking animal bends the tail forward over the body and stings with its tail spine.	A powerful neurotoxin, more toxic than most snake venoms but produced in smaller quantities, which depresses the activity of the nervous system.
Black widow spider *Latrodectus mactans*	Spiders have paired fangs and venom glands behind the mouth. This dangerous species can penetrate human skin about 0.4 mm.	Black widow venom is a neurotoxin, and may cause damage to the kidneys, liver and other organs.
Colombian arrow-poison frog *Phyllobates aurotaenia*	Poison is produced by special skin glands in many frogs and toads and is purely defensive. There is no apparatus for injecting it.	The venom extracted from poison-arrow frogs has been named 'batracho-toxin'. It is a very toxic steroidal alkaloid, and affects the heart directly.
Adder *Vipera berus*	These three animals are all front-fanged poisonous snakes with highly modified hollow teeth which fold into grooves in the mouth when not in use and spring erect when the nake strikes, to inject the venom produced in modified salivary glands.	
Gaboon viper *Bitis gabonica*		Snake venoms are complex mixtures of substances, including neurotoxins (particularly important in cobras and sea snakes), cardiotoxins (acting on the heart, and important in the Indian cobra) and haemotoxins, which destroy blood cells (the most important component of viper and adder venom).
Horned viper *Cerastes cerastes*		
Yellow-bellied sea snake *Pelamis platurus*	Sea snakes and cobras have relatively short, fixed, hollow front fangs. The venom is produced in modified salivary glands. Spitting cobras can inject venom in the usual way, but often raise the head and violently spit a stream of venom towards the eyes of the victim.	
Indian cobra *Naja naja*		

Natural use of venom	Impact on man
Stinging tentacles envelop small fish which are instantly paralysed by the venom and then eaten.	Probably the most venomous marine organism known, this Australian jellyfish can kill a man in 3-8 minutes. An antiserum called 'sea-wasp antivenene' has been developed which is of considerable value, especially if it can be given quickly. Victims are stung while swimming.
A tooth is projected into the prey, small fish in this cone. The paralysed fish is then engulfed by the muscular proboscis and digested.	This mollusc has been responsible for 4 or 5 fatal stingings, and others of considerable severity. The victims have been shell-collectors attracted by its beauty. Numbness and pain are followed by paralysis and difficulty in breathing.
The octopus feeds mainly on crabs and shrimps which the poisonous bite quickly subdues.	This octopus has been frequently handled with no ill effects; if it is roused and bites it causes local numbness and swelling, followed by paralysis of voluntary muscles. Most victims recover completely in 1–4 days but one man is known to have died. The species lives on the coasts of Australia.
Scorpions catch their prey with paired pincers; if it resists the sting is brought forward and the poison used. Male scorpions may feint with their stingers in combat, but do not usually sting each other.	This is one of two species which between them bite about 1,000 adults and children each year in the south-west USA. There is about one fatality a year children are more at risk. A specific antivenin is available.
All spiders kill their prey by injecting venom with their fangs. The black widow spins a web to trap small locusts, ants and flies before biting them.	This notorious spider has taken up a suburban way of life in America; it is abundant in such localities as outdoor privies and under cane seats in airport lounges. Perhaps 500 people a year are bitten with a fatality rate of less than 1%. An antiserum is available.
The poison of frogs is a defence against predators; a carnivorous mammal that picks one up will quickly drop it, or be seriously affected by the toxin. The bright warning colours of this frog make it unlikely the same predator will attack one twice.	The poison from this frog is extracted from them, by inducing stress, by the Choco Indians of Colombia. The extract is then used to poison blow-pipe darts, which are used for hunting monkeys and other small game.
	The only British species of poisonous snake, the adder or viper is of a retiring nature and rarely bites man. There have been about 12 snakebite fatalities in Britain this century. Viper antiserum is available, but is usually only given to children as some people are allergic to it.
Attacking prey and for defence.	Fortunately this largest of the vipers is a relatively good-natured and shy species. When it has bitten humans the effects have been particularly severe, with few survivors. Lives in wet forest regions of Africa.
Attacking prey and for defence.	A common, aggressive and highly toxic species of desert areas in the Middle East.
Attacking prey and for defence.	This sea snake is widespread in the Pacific Ocean. The usual victims of its bites are fishermen who catch them accidentally. In Malaysia in a population of 40,000 there were 144 bites in 11 years, 41 fatal.
Attacking prey and for defence.	India has one of the highest rates of snake bite in the world and the two most dangerous snakes are the small Krait and the Cobra, the snake charmer's snake. Perhaps 70,000 people are bitten each year, and of those 5% will probably die.

THE FAST
—AND—
THE SLOW

A small pod of bottle-nosed dolphins *(Tursiops truncatus)*. Dolphins have been known to reach speeds of over 50 kilometres per hour (31mph).

The Cheetah

Built for speed

Far right: A cheetah (*Acinonyx jubatus*) lashes out to defend its kill.

On the wide grassland plains of Africa the cheetah stalks its prey slowly, carefully. It moves upwind so its own scent will not betray it to the herd of gazelles that are grazing unsuspecting, but always alert, a few hundred metres from the cat. There is little cover, but the hunter uses such vegetation as there is, freezing to immobility at the slightest hint of nervousness from its quarry. It takes fully fifteen minutes to cover 25 metres (80 feet).

Its behaviour does not betray the fact, but as soon as it is close enough to examine the individual gazelles it selects its victim. This is the animal it will attack. The others are safe – though they do not know it.

A rustle, a movement of the grass that could not have been caused by the wind, a glimpse of black spots on a tawny yellow coat or of still, deep yellow eyes in a face whose dark lines, on either side of the muzzle, give the cheetah a careworn look, and the gazelles are warned. They do not wait for confirmation of their suspicion, but begin to run, all together. For the one among them that has been chosen, the warning came too late. The cheetah can now out-run a gazelle.

Speed and Hunting Tactics

All the large mammals of the African plains are capable of a fair turn of speed. Even the

Right: Though not known for its climbing ability, the cheetah will scramble up the low bough of a tree in order to gauge the movements of its prey.

Distribution of the best-known cats

Canadian lynx
Jaguar

Bobcat   Jaguarundi   Spanish lynx   Caracal   Clouded leopard

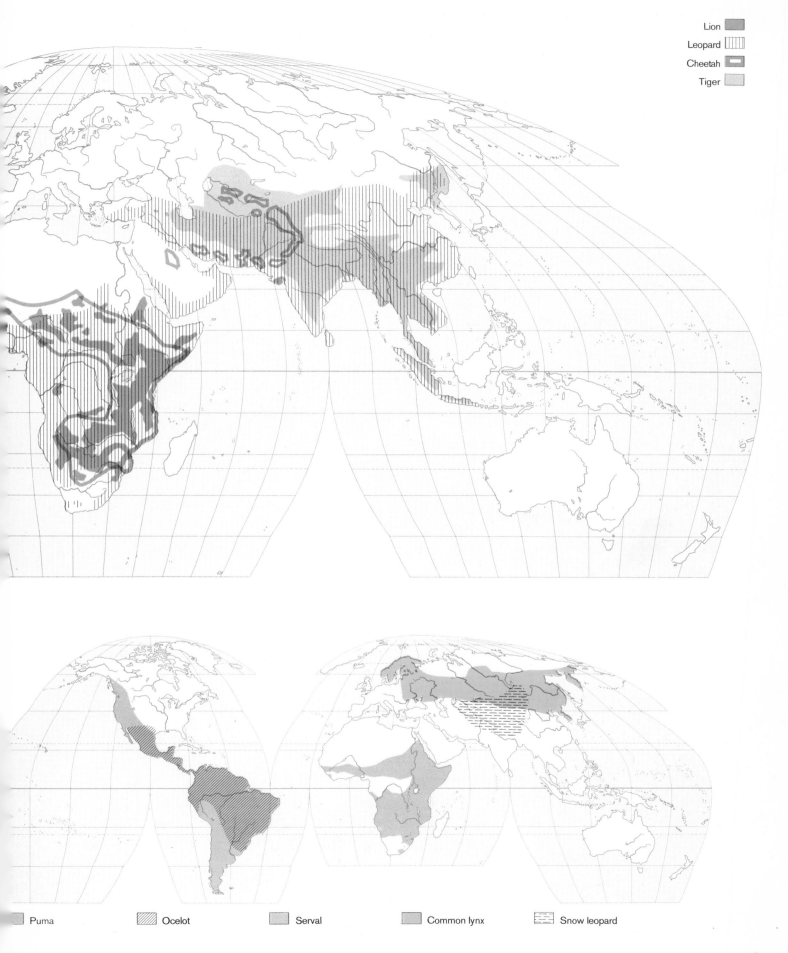

Lion

Leopard

Cheetah

Tiger

Puma   Ocelot   Serval   Common lynx   Snow leopard

Far right: Cheetah cubs stay with their mother for up to sixteen months.

Below: Cheetah on the alert.
Middle: Stalking.
Bottom: About to take a young gazelle.

rhinoceros and the elephant, slowest of them all, can run at 40 kilometres per hour (25 m.p.h.) when the need arises. The zebra has a top speed of about 65 km/h (40 m.p.h.) and the gazelle is capable of 80 km/h (50 m.p.h.). In a landscape that affords little cover beneath which a predator can move unseen until it is close enough for a single sprint and leap to bring down its prey, the race is to the swift.

The hunters of the plains must be able to run down animals they can obtain in no other way. True, the lions hunt as a team, the males driving the prey into an ambush prepared by the females, and the dogs and hyenas run as a pack, but even for them success depends either on cunning – as with the lions – or endurance – as with the dogs and hyenas – or on sheer speed. The cheetah relies on speed and it can overtake any land mammal in the world. A cheetah can run at 112 km/h (70 m.p.h.). What is more, it can reach this speed from a standing start in less than two seconds.

Its entire body is built for speed. It has long legs and a long back that can bend almost double. This gives the cheetah a huge stride. By arching its back and bringing its hind legs forward, then pushing hard with its feet and stretching its back, it can proceed in a series of bounds. It cannot sustain this great speed, however. The expenditure of so much energy makes great demands on the oxygen supply in its blood and after a sprint of 500 metres (nearly 550 yards) it must rest. After it has made its kill the cheetah must spend several minutes recuperating before it can feed.

The kill, too, makes great demands on the cheetah, for its technique differs from that of other cats. They rely on a single lethal bite to the back of the neck that severs the spinal nerve cord and causes instant death. The cheetah, more like a dog, attacks the throat and suffocates its victim. To achieve this it must grip for seven or eight minutes.

Usually alone or with cubs, but sometimes hunting in small family groups, the cheetah takes gazelles in preference to most other animals. They account for up to ninety per cent of its prey, probably because they are of precisely the right size for the cheetah attack strategy. This involves chasing the victim until the cheetah is close enough to deliver a blow that brings the gazelle to the ground. If the victim regains its feet it is struck again, as many times as are needed, until the cheetah is able to bite its throat.

The cheetah (*Acinonyx jubatus*) differs from other cats in several ways, the most marked of which is its possession of claws that cannot be retracted. It is not difficult to tame and has been used in hunting for centuries. The two to four cubs that comprise a cheetah litter have grey manes which they do not begin to lose until they are two months old, when adult markings begin to appear.

The Snail

Mollusc with a mobile home

The snail is a slow-moving, diminutive creature which seems almost oblivious to its surroundings, and whose lifestyle is generally thought to be as dull as the colouring of its shell. Compared to their more flamboyant marine relatives land snails are indeed quite conservative in form, but closer examination of their physiology and behaviour reveals some most interesting features.

Right: The ram's horn snail (*Planorbis carinatus*) can live in extremely stagnant water.

Most snails transport themselves by a succession of muscular ripples of their foot. This can be best observed by watching a snail making its way up the other side of a sheet of plate glass. There may be up to ten dark bands on the foot of the snail, all moving forward at the same time. To aid movement a snail secretes from the front end of its foot; as soon as the mucus emerges it is spread out as a smooth bed of lubricating fluid along which the snail moves. This enables a snail to cross all types of terrain and is especially useful when crossing patches of hard cracked soil. Snails can even walk over a razor blade with this protective coating.

There is no need for a snail to hurry as it investigates the state of decomposition and edibility of the vegetation it encounters. One of the slowest species found in Britain is the round-mouthed snail *Pomatias elegans* which moves at the rate of 2.5 centimetres (1 inch) per minute, equivalent to 1 kilometre every 27 days (1 mile in 42 days). In this case the sole of the foot is divided along its length down the middle into two functional parts. One half is raised at the front, contracted at the back, extended forward and then put down again. The other half is then lifted and goes through a similar sequence of movements. The whole painfully slow progression has been likened to that of two feet shuffling forwards in a sack. The common Kentish snail (*Monachia cantiana*) is carried forward by normal waves of muscular contraction of the foot at a rate of 7.5 centimetres (3 inches) a minute, equivalent to 1 kilometre every 9 days (1 mile in 14 days). Water snails progress similarly, by laying down their mucus trail they can even glide under the surface film. Some very small snails actually paddle along their mucus carpet by using thousands of minute hairlike processes called cilia as oars.

Galloping Snails

Some species of *Helix,* for example the marine snail *H. dupetithoursi,* can change gear and literally gallop along in what is known as a 'loping' gait. This means of progression starts by the head and front part of the body being lifted off the ground whilst still being propelled forward by the hind part of the body. The head is then lowered to the ground forming an arch and muscular waves now pass from the front of the animal backwards. This type of locomotion is mechanically similar in principle to earthworms and certain species of nemertines commonly known as bootlace worms. The 'loping' gait with the development of large vertical waves moving down the body, has the advantage of reducing to a minimum the frictional resistance of the snail's foot with the ground.

Reproduction and Physiology

For such apparently simple-looking animals, snails have a surprisingly complex mass of internal organs, many of which are concerned with reproduction. The majority of snails are hermaphrodite and have the full complement of both male and female sex organs plus additional ones involved in sexual stimulation and protection of the eggs. Courtship is generally a lengthy process; in the edible or Roman snail a pair will raise their soles off the ground and come together with gentle swaying movements and actively caress each other with their greatly expanded head tentacles. As they reach a climax both eject calcareous darts which penetrate deep into the body of their partner. This dramatic mutual exchange stimulates copulation in which the penis of each snail is inserted into the vagina of the other. It takes four weeks between fertilization and

laying of the eggs and by this time each egg is protected by a thick albumen coating and a shell. The eggs are buried in a shallow burrow which is covered up to protect them from predators and from drying out.

Drying out is one of the major problems for snails and although they are found in a wide variety of habitats including deserts they become dormant in dry conditions, either hibernating as they do in Britain or aestivating during the hottest summer months. In both cases snails resist desiccation by sealing the entrance of their shell with a thick temporary seal of dry mucus known as the epiphragm, which is composed mainly of calcium phosphate. Clusters of garden snails (*Helix aspersa*)

Right: The great pond snail (*Lymnaea stagnalis*) feeds on detritus, plants, and dead animals.

are often found in the winter underneath a flowerpot or in a similar corner with their bodies cemented together by the accumulation of dried mucus. Along the Mediterranean similar groups of snails are found, but these differ in being generally of a brilliant chalky white (which efficiently reflects the heat), and in being clustered at the top of a dried stalk of grass to take advantage of the cooling influence found within a metre or so of the ground. During its winter sleep a snail's metabolic rate sinks even lower; the heart beat dropping from the normal 10–13 to 4–6 beats a minute.

A snail feeds using its radula, an organ equipped with many thousands of small teeth which is everted through the mouth and serves as a broad file to rasp away at plant tissues. Water snails crunch up small plants and scrape algae off larger plants; microscopic food is also collected by trapping it on a stream of mucus which is wafted towards the radula and mouth by the concerted action of numerous cilia. There are both slugs and snails of carnivorous habits. One particular species found in New Zealand is said to attack quite large earthworms which it spears with its radula and then draws into its mouth like a strip of spaghetti.

Predators

Despite their shells into which they can quickly retreat snails, like most animals, have their enemies. The song thrush is a notorious snail basher. Gripping the rim of a shell with its beak it will smash it repeatedly against a stone to get at the succulent contents. A thrush often uses the same anvil regularly, which is often littered with shell fragments. The other main snail hunter, the glow-worm (*Lampyris noctluca*), of which the larva is especially voracious, has a more subtle approach. The glow-worm first paralyses its victim by injecting a poison into the soft mantle of the snail and then proceeds to suck up the body tissues which are broken down by this substance into a semi-liquid mass. The French naturalist J. H. Fabre described the whole event in lurid detail.

Then the hunter's weapon is drawn. . . It consists of two mandibles bent back powerfully into a hook, very sharp and as thin as a hair. The microscope reveals the presence of a slender groove running throughout the length. And that is all.

The insect repeatedly taps the Snail's mantle with its instrument. It all happens with such gentleness as to suggest kisses rather than bites.

Above left: The giant snail (*Achatina achatina*) of equatorial Africa attains a length of over 30 centimetres (1 foot).

Above: Three colour forms of the white-lipped snail (*Cepaea hortensis*).

Right: The banded snail (*Cepaea nemoralis*).

Far right: Snail eggs receive double protection against drying out – they are enclosed in a thick gelatinous shell and are laid in batches in a shallow burrow which is then filled in by the parent.

Right: Cross copulation of the Roman snail (*Helix pomatia*).

The Peregrine Falcon

Daring killer of the skies

The barred wings of a peregrine (*Falco peregrinus*) quartering territory that it has chosen as its hunting ground is a sight that thrills observers in all the world's continents and many of its islands. It is a thoroughly cosmopolitan bird, rejecting large tracts of the steamy jungles of South America and the wastes of Antarctica, but living well elsewhere: an indication of the success of its lifestyle and adaptability to varied environments. Species of small peregrines hunt the fast-flying passerines of the deserts, and the large northern falcons will select even heron or small bustard and ravens as prey – an ambitious choice for a falcon of only 600–800 grams (21–28 ounces).

The peregrine's stubby body, pointed wings, close plumage and short wedge of a tail combine to make it one of the world's most enthralling aerial acrobats. Like most predators, the peregrine flies with short, fairly rapid strokes of its wings, occasionally making use of thermals to lift it in graceful glides as it searches for prey. In level flight it is not particularly fast. It cannot match the speed of its commonest victim, the pigeon. The peregrine's flapping flight bears it along at between 75 and 100 kilometres per hour (47–63 m.p.h.); but in attack, its technique changes, and its speed becomes deadly.

Experience seems to have taught the peregrine the right approach to its swift adversary. The pigeon's eyesight is good, but it suffers from a blind spot behind its head. To the front, above and directly below, the pigeon can spot attackers quickly, and put on a turn of speed that leaves most of them behind. The peregrine makes its attack from behind, striking home before the pigeon can accelerate and make a run for safety.

The peregrine is one of the few avian predators which hunts a quarry while it is in flight: a feat that requires microsecond timing, strength and daring. The falcon generally climbs to a good height from which it can survey its hunting ground. Its dark eyes, so dark that it is hard to distinguish between pupil and iris, have between four and eight times the power of resolution possessed by human eyes. They see in full colour, and the falcon's binocular vision is excellent. Not for nothing is good sight described as 'hawk-eyed'. Little that moves will escape the peregrine's notice. As it examines the space below, and sweeps the ground for any movement that might indicate prey, the falcon will pause, achieving this mid-air stop by tilting its body towards the vertical and fanning the breeze with its wings. Hovering above its hunting grounds, the peregrine avoids transmitting the danger signal of its movement through the air, and takes the chance of a more minute examination of the ground below.

The Stoop

When the peregrine spots its prey winging across the space below, it unhurriedly works its way onto a path that will intercept. A pigeon which fails to notice the speck floating from thermal to thermal high above its wake will often fly a steady course unaware of the falcon's fast shallow dive that is closing on its path. The peregrine's final tactic is almost irresistible. It folds its wings back until the tips fall into line alongside the broad tail, shrugs its shoulders and, with a few final strokes of the backward extended wings, drives itself into a

Below: This portrait of the peregrine falcon (*Falco peregrinus*) shows the tooth on the upper mandible which is used to break the prey's neck once it has been struck down by the long talons.

powered dive that often exceeds 300 kilometres per hour (188 m.p.h.).

As the dive sweeps the peregrine a few metres behind the pigeon's tail, the falcon's wings spread out to flick its body into a vertical plane, while its speed carries it almost level with the pigeon's body. In this way, the peregrine makes full use of its victim's blind spot. At the last moment, as the attack is pressed home, the peregrine tilts its pelvis forward and drives its powerful feet forward and down to stamp its talons deep into the pigeon's body. The shock of the impact at high speed, the final punch of the feet, the penetration of the sharp talons, their grip – as powerful as the 'vigorous handshake of a strong man' – combines to produce massive shock that will kill most birds outright.

The method of attack makes the fullest and best use of the peregrine's physique and senses. The approach from height that the bird usually adopts is notoriously difficult to detect – ask any Second World War fighter pilot. The stoop (dive) is a tactic performed better, and faster, by the peregrine than by any other bird. Some authorities estimate its speed to be a little more than 250 kilometres per hour (156 m.p.h.), but recently improved measuring techniques suggest that this is far too low. There are ornithologists who now believe the falcon stoops at up to 440 kilometres per hour (275 m.p.h.). Despite the pressures of this speed, the bird continues to breathe, and can judge its attack perfectly. The final levelling out makes it possible for the peregrine to stamp home its assault without damaging itself. If the stoop were driven straight home, the attacker's weight bearing down directly on the victim, it is inevitable that the falcon's legs would break. The flattening of the stoop as the bird's legs are held flexed allow it to arrive at the point of impact at high speed and to deliver a powerful, though often a glancing, blow along the back of its prey. This is usually sufficient to kill a pigeon or at worst knock it to the ground where the peregrine can then dispatch it.

The kill is not always made in the air. Sometimes, the falcon must carry a large, injured and shocked prey to the ground, and use its strong curved bill to break the bird's neck. Once the bird is dead, the peregrine generally severs its head, plucks the feathers from the breast, and opens the belly. It often picks the head clean of food and eats the viscera

before carrying the lightened load to a more convenient eating place. In season, it will fly the carcase back to its mate and fledglings.

Family Life and Conservation

The brilliance of the peregrine's flight is not confined to its miraculous stoop. As a hint of spring warms the air, young tercels (males) look for falcons (females) to mate. Established pairs remain together for life, but the rituals of courtship are celebrated afresh each year as the mating season returns. A tercel's approach to his mate is cautious, even wary. She is substantially larger and more powerful than he is, and if he annoys her she may well push home a punishing attack. The male displays the beauty of his flight in magnificent circles near the chosen nest site. Most of the year the pair will be silent, but at mating time and when one approaches the other during the incubation of the eggs they utter loud screams or chittering cries rather like those of a kestrel.

The circling flight gradually becomes more acrobatic. A dipping, undulating pattern develops, sometimes breaking into a shallow dive, a swoop upward and another dive. When the exhilaration of the display seizes hold of the bird, he will extend the dives and

Above: The peregrine is favoured by falconers for its speed, strength and consistent performance. Flight is fast and made up of alternate wing-beats and gliding, with a nose-dive stoop on to the prey.

93

swoops into a figure-of-eight looping pendulum of immense grace. It is hard to avoid the belief that the birds feel a great sense of pleasure in this virtuoso performance.

When their pale buff eggs with strong rufous markings have been laid in the crude nest of sticks, the female usually does most of the work of guarding and incubating at first. Being the larger of the pair, she is better able to mount a formidable static defence. The tercel hunts for food for both of them and, later still, the hunter will bring back food for the young eyasses (hatchlings) as well.

When he returns to the nest after a successful kill, the hunter calls to warn his partner with a piercing scream. She rises from the nest and flies to meet him. High above the earth, she screams to him, and as she arcs below him he releases the half-plucked bird from his talons. As it drops away from his feet, the falcon below turns upside down for a moment to catch the prey with her feet before flying back to the scrape (nest). This extraordinary game of catch in mid-air is one of the most thrilling sights in nature. It demands an exquisite sensitivity to the buffeting winds and a minute sense of timing.

Decline

It is a sad fact that our chances of seeing peregrines displaying this supreme mastery of the air have grown fewer during this century. In Britain during the two World Wars, peregrines were treated as enemies. The governments of those days commanded that they should be exterminated because they feared the birds' skill in killing pigeons. In both wars homing pigeons played a part in transmitting intelligence from the European mainland to these islands, and the thought that any of these precious messages might fail to arrive led the war leaders to extend hostilities to the peregrines. They were shot on sight, even when seen far inland, and their nesting sites were destroyed, eggs confiscated and their young killed. Slowly the British population began to recover after the end of hostilities. Then another, and still more serious threat appeared.

In Europe and North America the numbers of peregrines began to decline to seriously low levels. Investigations showed that this predator was eating the organs of birds which had high concentrations of chlorinated hydrocarbons in them. The peregrines' eggs too showed these residues of the growing use of pesticides. The consequence of the birds' natural diet was that they laid infertile eggs, and the entire population was at risk. In the eastern United States, and southern Britain, the population of peregrines fell to disaster level. Fortunately, the species is a resilient breeder and, after the withdrawal of some of the most dangerous of the pesticides and their replacement with safer ones, the falcons show signs of restoring their populations, albeit very slowly.

In the United States captive breeding of the American peregrine (*Falco peregrinus anatum*) has been very successful and over two hundred birds have been so far transferred to the wild. Young peregrines recently released in large cities like New York and Washington have fared well on a diet of pigeons, starlings and sparrows. The main hazard facing these birds is the countryside where there is competition for cliff-face nest sites with another powerful predator, the great horned owl which has been known to attack and feed on young peregrines as they roost at night.

The Sloth
Slowcoach of the jungle

If the cheetah is the fastest of all mammals, the sloth is the slowest by far. Its top speed, which it cannot maintain for more than a few metres, never exceeds one kilometre per hour, but the sloth seldom rushes so fast. Everything about it is slow. It spends eighteen hours of each day fast asleep. It digests its food so slowly that it needs to urinate and defecate only once a week. The fastest thing it can do is to make a sweep with one of its long forelegs. Usually it does this in order to catch and pull toward itself a branch bearing the leaves on which it feeds. Its long, curved claws are sharp and could inflict a serious wound on an enemy, but the attack would be so slow that the victim would have ample time to avoid it.

Not only is the sloth the slowest of all mammals, it is also the dirtiest. Personal hygiene acquires a new meaning in its world. It never washes. Its long, stiff fur is usually damp, for it lives only in the rainforests of the American tropics, and while many animals harbour fleas or mites, sloth fur harbours algae: unicellular green plants. They give the animal a green appearance which helps to camouflage it. The algae do not live alone, for wherever nature supplies plants there will be some animal that considers them food. A moth lays its eggs in the fur of the sloth and the caterpillars graze the algae.

All the senses of the sloth are dimmed. Its sense of smell is better than ours, but not so good as that of most mammals. It does not see well, is almost deaf, and although capable of uttering a shrill cry, for most of the time it is silent. A largely nocturnal animal, it spends its few waking hours moving upside down at a leisurely pace through the trees, quietly eating leaves.

Survival of the Slowest
Life is easy for the sloth, and that is why it has survived. Long ago its ancestors lived on the ground, but the ground sloths are now extinct. Today this relative of the anteaters and armadillos is the most arboreal of all mammals. There is no enemy that can reach it high in the trees, and its body is well designed for its way of life. The forelimbs are much longer than the hind limbs – which is useful in climbing – and the large claws are hooks, by which it hangs securely and effortlessly. On the ground a sloth can move only by dragging itself along with its forelimbs and is quite incapable of defending itself against the jaguar that prowls the forest floor. In the trees it is so difficult to see that for many years people believed it was rare. In fact it is quite common.

Despite its reluctance to descend to the ground, once a week it must do so, for a sloth always urinates and defecates in a particular place. This curious habit, in an animal far from fastidious in other respects, may provide a clue to a mystery concerning sloths. How do they mate? Most sloths live alone except when a female has an infant hanging on to her fur – although occasionally two adults may be seen together. Their senses are so poor that there would seem to be no way in which males and females could locate one another and meet. The midden may provide the answer. It has a strong smell and it may be that a sloth can distinguish between the midden belonging to a male and that belonging to a female. If so, a sexually receptive female might be able to wait in some secure place close to a male midden until its owner appeared on his regular weekly visit.

The fact is that the behaviour of sloths has been little studied for the simple reason that few scientists have been willing to endure the tedium that this would involve.

Species
The family Bradypodidae, to which the sloths belong, consists of two genera and approximately seven species distributed through the forests of tropical America from Honduras south-eastwards to northern Argentina, Paraguay, and Brazil. The two genera are the three-toed (*Bradypus*) and the two-toed (*Choloepus*). As the names imply, the three-toed sloth, known in South America as the 'ai', has three toes and claws on each limb, and the two-toed, known as the 'unau', has two on the forelimbs and three on the hind limbs. Of the

two, it is the three-toed 'ai' that is the slower. There are other differences. The three-toed sloth has nine neck vertebrae, for example, which enables it to turn its head round through 270 degrees. The two-toed sloth has six or seven (most mammals have seven). Although the two-toed sloth has been known to live for more than twenty years in captivity, the three-toed sloth is much more difficult to keep in artificial surroundings. This is probably due to its selective diet – for it feeds almost exclusively on the leaves of the *Cecropia* tree, a giant relative of the temperate-zone mulberry.

To digest its bulky vegetable diet a sloth's stomach is equipped with several chambers based on a similar principle to the digestive system of ruminants. The first section of the stomach mechanically breaks down the food by the action of strong muscles, while the other sections secrete powerful enzymes. The stomach is always packed to capacity and may constitute up to 30 per cent of the body weight.

To summarize the slothful activities of this animal, the patient American naturalist William Beebe once followed a sloth continually for a week and found its routine consisted of 11 hours feeding, 18 of slow movement, 10 hours of rest and the remaining 129 hours of sleep. In keeping with this lazy lifestyle, sloths are attributed with a low intelligence.

Below: Mother and young of the three-toed sloth (*Bradypus tridactylus*), the slowest-moving land mammal.

The Killer Whale

Speed-king of the sea

The killer whale (*Orcinus orca*) is the fastest and most voracious marine mammal. When chasing prey it can reach a top speed of over 65 kilometres per hour (40 m.p.h.), but normally cruises at a more leisurely pace of 10–13 kilometres per hour (6–8 m.p.h.). The propulsive force is produced by the broad tail flukes which are thrashed up and down parallel to the water surface driven by powerful muscles situated just in front of the tail. When a killer whale is at the surface and prepares to sound it first points its head down and arches its back, the resulting wave of muscular contraction passing down the body to the tail flukes, when the energy is released with a great thrash of the tail which has been likened to the cracking of a mighty whip. By a rapid increase in the frequency of movement the tail flukes can propel a killer whale out of the water producing spectacular 8-metre (26-foot) leaps. A killer whale will often raise its head and body out of the water to investigate its surroundings and can 'tail walk' by leaping up until all that remains underwater are the flukes which are thrashed around vigorously to maintain the upright posture. Large bulls will slap the water with their extra-large paddle-shaped fins or may leap out of the water to crash back broadside on with a resounding splash which can be heard for up to 8 kilometres (5 miles).

The killer whale is a beautifully proportioned giant dolphin. The profile of the head runs in a smooth curve from the tip of the snout over the back in a continuous unbroken sweep. Apart from its distinctive black and white markings it is distinguished from other dolphins by its prominent dorsal fin which in adult bulls may project 2 metres (6½ feet) out of the water. In males this fin is triangular but is more slender and sickle-shaped in females. There is also a great disparity in size between the sexes, as an adult bull may grow up to 9 metres (nearly 30 feet) whereas the maximum length for a female is 6 metres (20 feet) though they are generally much smaller. However, even a small female can weigh nearly a tonne.

Killer whales have a cosmopolitan distribution and are found throughout the oceans from the tropics to the pack ice of the polar regions. They are the supreme marine predators and are the only members of the order Cetacea to take warm-blooded prey including other dolphins, whales, seals and seabirds though they are thought to feed mainly on large fish and squid. To tackle its prey a killer whale is equipped with strong plug-like teeth implanted in deep sockets, twenty to each jaw.

Hunting Techniques

Killer whales are social animals swimming in packs or pods of from three to forty individuals. Like the more familiar common dolphins they emit a large range of ultrasonic sounds used to locate prey and as a means of communication between members of the group. When hunting, killer whales are capable of sophisticated group manoeuvres. Off the coast of Baja California, Mexico, a pack of killer whales were seen to attack a large group of dolphins. They swam round the prey in ever-decreasing circles until the dolphins were packed close together. Then one killer rushed out from the group and snatched a dolphin in its jaws while the others continued circling; in this way, taking it in turn, they wiped out nearly all the dolphins. From the few authenticated reports of killers attacking larger prey such as baleen whales, their tactics appear to be to strike en masse, lunging at the whale's soft underparts. A young grey whale was killed and eaten by a pack of killers at the calving grounds off the Baja Peninsula, and though the mother was injured she managed to escape leaving a gory trail of bloody water. The minke whale is known to have been the victim of an attack off Vancouver Island. Killers tend to concentrate on sick or young individuals and are known to tear chunks of flesh and blubber off recently harpooned whales.

In northern waters killer whales are renowned for their attacks on seals marooned on ice floes. A group of killers will dive down and then surface, ramming the ice floe with their powerful heads until the seal is knocked off into the water. Similar behaviour was witnessed by members of Scott's Antarctic expedition in

Above: With a mighty thrash of its tail flukes a bull killer whale leaps above the surface.

1905 when two husky dogs were leashed to the expedition ship on an ice floe that was encircled by killer whales. As a photographer drew near to photograph them the killers disappeared under the water but a few minutes later rose up under the ice, fragmenting the floe into several parts. Fortunately the dogs survived this attack. It is, however, a fact that dogs straying too near the sides of aquaria containing killer whales have, on occasion, been snatched and eaten.

Man and the Killer

Despite its record for tackling large mammalian prey there is no evidence of a killer whale ever deliberately attacking a man. Fishermen's boats have been rammed and overturned by them but the occupants, though terrified, have been left unscathed.

Since the first killer whale was taken into captivity in 1964 they have been widely used for training and research and have been shown to be equally as intelligent as the more familiar bottle-nosed dolphin. In some aquaria notably in Canada and the United States trainers ride on their backs, make them jump through hoops and fearlessly place their heads and arms into their huge gaping jaws.

Members of the conservation organization Project Jonah in Canada have paddled in

FAMILY DELPHINIDAE

Sub-Family Orcinae
Pseudorca crassidens
The 'false' killer whale which has been found stranded from Tasmania to Kiel.

Orcinus orca
The killer whale, a cosmopolitan sea mammal, not uncommon in seas of Britain and the US.

Orcaella brevirostris
The Irrawaddy dolphin found in the Irrawaddy River, Burma up to 900 miles from the sea. Also in the Bay of Bengal.

Globicephala melaena
The pilot or caa'ing whale, commercially hunted in the North.

Globicephala macrorhyncha
The pilot whale of the tropical seas, found in the Indian and Atlantic Oceans.

Feresa attenuata
The pigmy killer whale found in the Japanese seas and Senegal.

Sub-Family Lissodelphinae
Lissodelphis peroni
The right whale dolphin, found in Southern seas.

Lissodelphis borealis
The right whale dolphin of the North Pacific.

Sub-Family Cephalorhynchinae
Cephalorhynchus commersoni
The piebald porpoise or 'Commerson's dolphin' found in Southern seas.

Sub-Family Delphininae-Lagenorhynchus
Lagenorhynchus albirostris
The white-beaked dolphin found in the North Atlantic.

Lagenorhynchus acutus
The white-sided dolphin, also found in the North Atlantic.

Lagenorhynchus obscurus
The dusky dolphin of the South Pacific.

Lagenorhynchus australis
The dolphin of the Southern oceans.

Grampus griseus
Risso's dolphin which is cosmopolitan and quite common in British waters.

Tursiops truncatus
The bottle-nosed dolphin common in British seas and off the East coast of the United States.

Tursiops gilli
A bottle-nosed dolphin of the North Pacific.

Tursiops aduncus
A heavily pigmented tropical dolphin found in the South Pacific.

Delphinus delphis
The common dolphin, cosmopolitan in its movements and found in the Atlantic and the Mediterranean.

Stenella
The frequently spotted oceanic group of dolphins.

Right: In captivity the killer whale is gentle and has never been known to attack man.

Left: A combination of streamlined body shape and powerful musculature makes the killer whale the fastest marine mammal over short distances.

canoes among pods of killer whales that regularly visit the coastal waters off British Columbia. These are young people who in a world of high technology find a certain solace for their primitive instincts in mingling with these cetaceans. Apparently the whales respond to music played to them by the whale enthusiasts. Paul Spong, Director of Project Jonah in Canada, wrote:

> We often play music to the whales, for we feel their interest in appreciation of it. Sometimes, particularly on a still night, a pod or part of a pod, or perhaps just a single whale, will hover off-shore for an hour or more tuning into the music. Sometimes they seem to join in. . . .

Survival Games

Technically speaking, killer whales are not considered by conservationists to be an endangered species. Although the small-whale fisheries of Norway and Japan may very well have reduced stocks of the animal along their own coasts since the early 1950s, the comparatively small size of the killer whale has protected it from the appalling ravages of the world's whaling industry. Man has, however, damaged stocks of the killer whale by purely amateur cruelty. In the World Conservation Yearbook in 1978, the biologist Erich Hoyt reported that these fascinating mammals were very frequently shot by fishermen and other boaters from a sense of boredom or misplaced hatred. Fishermen, in particular, felt the killer whale was fair game because of its supposedly 'bad' reputation and enormous appetite for fish. Fortunately, by the late 1970s, these shootings decreased in number as the captive killer whales in North American leisure aquaria rapidly convinced the public of their great charm, playfulness, astonishing intelligence and, needless to say, the fact that as a species, they are no danger to man.

Above: The killer whale is known to isolate seals on an ice floe and repeatedly lunge at the ice from underneath until the ice breaks and the seals topple into the water.

HOW ANIMALS MOVE

Name	Method of locomotion	Speed
Cheetah *Acionyx jubatus*	Four-legged gallop when moving at speed, each stride 5-6 times the length of the body from chest to rump. The maximum speed can only be maintained for short bursts, up to about 500 m.	Maximum 110 km/hour
Three-toed sloth *Bradypus tridactylus*	In trees, sloths spend most time upside down, suspended from all four limbs, pulling themselves slowly along. They rarely descend to the ground, where they cannot stand or walk, but reach forward a limb for a toehold and pull themselves along.	Ground speed 0·11–0·15 km/hour Speed in tree 2·19 km/hour
Great grey kangaroo *Macropus giganteus*	Leaps on the two enormously developed hind legs, with the long and heavy tail counterbalancing the body.	Maximum 48 km/hour Maximum length jumped: 13·5 metres
Killer whale *Orcinus orca*	Propelled by up and down movements of the large horizontal tail flukes. Killer whales often leap 1·5 m clear of the water while swimming, and cover distances of 12–13.5 m in these leaps.	Maximum 57 km/hour Cruising 13 km/hour
Peregrine falcon *Falco peregrinus*	Narrow curved wings are used for rapid flight. The greatest speeds are reached when the falcon stoops, that is dives with half-closed wings.	Very controversial. A recent authority quotes: maximum level flight: 100 km/hour maximum stooping speed: 270 km/hour
Spine-tailed swift *Chaetura caudacuta*	The very narrow curved wings and forked curved tail are adaptations for the swift's almost completely aerial existence.	Maximum air speed 170 km/hour
Kiwi *Apteryx australis*	A flightless bird, the kiwi has strong but short legs on which it walks around, usually quite slowly, though it will run to cover if disturbed.	Not recorded
Sailfish *Istiophorus platypterus*	Moves by strong side to side movements of the streamlined body, helped by the extra strength of the spine which is provided with more interlocking joints between the vertebrae than most fish.	Maximum 100 km/hour
Dwarf seahorse *Hippocampus zosterae*	Has an armoured inflexible body which hangs upright in the water and is propelled by the movements of the tiny dorsal fin.	Maximum 0·016 km/hour
Dragonfly *Anax parthenope*	Flies with two pairs of long slender wings, the narrow cylindrical body presenting little resistance to the air.	Maximum recorded 28·57 km/hour Probable maximum 35 km /hour
Common snail Helix aspersa	The snail has a single large foot on which it glides over the ground by means of waves of muscle contractions lubricated by a trail of slime.	7·5 cm/minute
Squid *Onychoteuthis banksii*	At speed, squid swim by contracting the bag-like mantle which surrounds the body and forcing seawater out through a muscular funnel, a type of natural jet propulsion.	Maximum 50 km/hour

Feeding habits	Other points of interest
Stalks small antelopes, hare etc. to as close a point as possible, and then unleashes explosive final pursuit.	The fastest land mammal over short distances. Cheetahs live in open country in Africa and S. Asia.
Browses on leaves, tender twigs and buds.	Not only the slowest-moving mammal, all the body processes are very slow, food taking several days to pass along the gut. Sloths live in the rain forests of South America.
Feeds mainly by night on grass and other plants. Wanders in search of good grazing and water, though individuals can do without water for some weeks.	Probably some of the smaller kinds of kangaroo may move faster than the Great Grey Kangaroo, but its leaping ability and stamina are outstanding.
Chases and kills dolphins, seals and young walrus; killer whales hunt in packs and will attack and eat the larger baleen whales.	Killer whales are powerful swimmers, and can smash through ice a metre thick. They make longer migrations than any other whales, travelling through all the world's oceans.
Feeds on many kinds of small to medium sized bird, in Britain especially on pigeons. Peregrines have been known to kill geese and buzzards. The prey is attacked while it is in flight.	Probably the fastest bird of prey in level flight and diving. Peregrine falcons suffered a near-catastrophic population decline in the early 1960s, but have recovered in Britain to 75% of the pre-war numbers.
Catches many kinds of insects on the wing.	The fastest of birds in level flight. Swifts have difficulty taking off from the ground and drink on the wing flying low over lakes, and bathe on the wing, in rain storms.
Feeds at night on earthworms, which it locates when near the surface by scent and sound.	Amongst the slowest moving birds.
Overtakes and devours smaller fish and squid.	Probably the fastest of sea fish. The sail-like dorsal fin is folded into a groove when moving at speed to reduce water resistance.
Seahorses have very small rounded mouths. They feed on various kinds of crustaceans (shrimp-like animals), being limited to the tiny ones which are all they can swallow.	The slowest known fish, sea-horses drift in beds of sea-weed or sea-grass, holding on to vegetation with prehensile tails.
Dragonflies catch and devour smaller insects while on the wing, using their exceptionally good binocular vision to pinpoint the prey.	Dragonflies, hornets and hawk moths are the fastest of insect fliers, but none has been really accurately timed.
Snails graze while crawling slowly over plant material. The sandpaper-like tongue is protruded through the mouth and rasps away the plant surface.	The garden snail though slow-moving can pull 90 times its own weight when crawling on the level.
Squid prey on fish which they swim after and then catch with their suckered tentacles. In this species some tentacles are modified into hooks to grasp the most slippery fish.	This distant relative of the snail is the fastest swimmer of all invertebrates, and has extremely well-developed eyes to assist its hunting.

THE SOCIABLE
— AND —
THE SOLITARY

Six weeks after hatching young Emperor penguins
(Aptenodytes forsteri) gather together in crèches
while the parents hunt for food.

The Cuckoo

Rogue minstrel of spring

Right: A female common cuckoo (*Cuculus canorus*) caught in the act of removing the egg of a reed warbler before substituting her own.

There is nothing more evocative of spring than the clear repetitive two-note call of the common cuckoo (*Cuculus canorus*). For all its loud announcements the cuckoo is a most difficult bird to get close to. Invariably as you approach a cuckoo in a dense stand of trees or at the edge of a wood, the call suddenly stops and you will be lucky to catch a glimpse of a streamlined silhouette gliding away on long hawk-like wings.

The cuckoo flies over here from Africa at the same time as other smaller migrant birds, many of which are destined to become foster parents of its young. Most cuckoos reach our shores by the second week of April, 14 April being the day when the first call is most frequently heard. The typical male call note is used to advertise his territory and also to attract female birds. A receptive female will ruffle her feathers and spread out her tail in the form of a fan and the actual mating often takes place on the ground. After the brief courtship and mating the sexes part and lead solitary lives for there is no need for a pair-bond. Once her eggs have been fertilized the female prepares for her main task of depositing them in the nests of other birds. Cuckoos return to the area where they were initially reared and during the several weeks it takes for her eggs to develop, the female bird scrutinizes the nest-building activities of the songbirds in her vicinity, paying particular attention to the one species in whose nest she was previously reared.

From Egg to Fledgling

From now on the story is an extraordinary one – the female cuckoo usually 'strikes' in the afternoon when there is a general lull amongst the bird community. Flying swift and low the cuckoo lands on the nest and by means of her extensible cloaca lays a single egg into the nest of the host bird. Often the nest is so small that the egg must actually be dropped into it, and so to prevent breakage the egg is unusually thick-shelled. Egg laying is an extremely fast operation and it is a mere ten seconds before the female flies off to find another nest. It has been suggested that the

hawk-like shape and flight of the cuckoo as it approaches the nest causes the birds to vacate and hide for the brief period required for laying. After the substitute egg is deposited the cuckoo removes one of the host bird's eggs and the deed is done! Although the eggs of the common cuckoo are slightly larger than those of the host, they match them in both colour and markings. Cuckoos' eggs show a large variation in overall colour from white, blue, blue-green, yellow-green to red-brown and chocolate brown. One of the best examples of egg mimicry is demonstrated by the Japanese race of the common cuckoo which lays its eggs mainly in the nests of buntings. This cuckoo's egg is a marvellous replica having all the typical scribblings, irregular streaks and spots of the bunting's egg.

Each female cuckoo is however restricted in her choice of hosts to one or at most two other bird species. The one anomaly seems to be the hedge sparrow where the cuckoo's egg often does not bear any resemblance to its own egg. Nevertheless it is not rejected and perhaps the dissimilar egg hue is due to the hedge sparrow being a relatively recent host. In England the cuckoo regularly parasitizes tree and meadow pipits, wagtails, reed warblers and robins. Throughout its summer range the cuckoo has been known to lay its eggs in the nests of nearly 300 species of birds.

Despite the innate cunning of the cuckoo, its eggs are sometimes detected and the foster bird may abandon its brood and build another nest, but the cuckoo is persistent and will repeatedly try and deposit its eggs in the re-made nest. If a host bird's brood is too far advanced in development, a cuckoo will destroy the nest and contents and force the bird to rebuild. Though cuckoos are reported to lay an average of ten to twelve eggs their laying capacity is much greater if a larger number of suitable nests can be found. This fact was proved by the ornithologist E. P. Chance who removed the nests of a meadow pipit just as they were completed and filled with a cuckoo's egg. When the new nest was built and the cuckoo motivated to lay again, the

nest was again removed. This was repeated over a two-month period during which time the pipit made a total of twenty-five nests each of which contained a cuckoo's egg.

When the cuckoo's egg is accepted into the nest unnoticed, it is brooded by the foster parents along with the rest of the clutch, and develops at a much faster rate than the other eggs, having an incubation period of only twelve days. As soon as it hatches out, the large naked chick shows a remarkable urge to eject everything else out of the nest. Using its strong stumpy wings it lifts an egg or nestling onto its broad hollow back, then, supported by its feet and the front of its head, it backs up to the edge of the nest and heaves its victim out. This is repeated until the whole clutch has been ousted and the cuckoo remains the only occupant. Apparently the parent birds pay little attention either to the eggs or the pathetic-looking nestlings which often remain stranded on the rim of the nest. Probably the adult birds are sufficiently distracted by the young cuckoo whose huge gape of a mouth with its bright orange lining serves as a super-stimulus to them.

Five days after hatching the young cuckoo's feather quills appear. For the next three weeks it is fed continually by both foster parents and rapidly gains weight until it is some forty or fifty times heavier than when it was hatched. One of the pioneers of cuckoo behaviour, F. Jourdain, wrote of the young cuckoo: 'Nothing more unlike the nestling of most ordinary fosterers can well be imagined than the great brown bird with huge orange mouth and stumpy tail which sits on the crushed remains of what was once a nest and keeps up its penetrating call, which . . . is quite unlike the ordinary notes of their young.' Even when the youngster has left the nest the birds continue feeding it for a few weeks until it is able to catch its own food. There have been cases of other birds apart from the foster parents feeding young cuckoos, apparently unable to resist stuffing food into the ever-open mouth.

Cuckoos appear to specialize in eating caterpillars that other birds shun – especially those with long irritating hairs and ones with bright warning colours which advertise their distastefulness. When tackling a poisonous caterpillar a cuckoo will knead it with its bill, then, grasping one end, will vigorously shake it around, expelling the noxious body contents before it swallows the remaining part. The

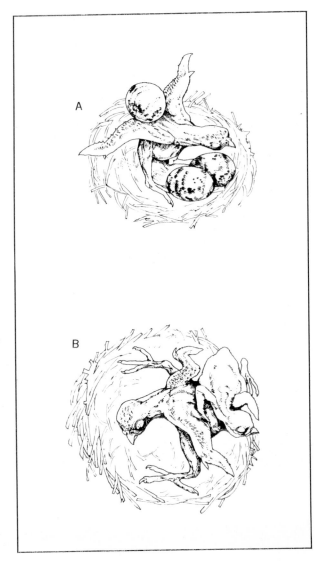

Above: A young cuckoo begs for food while the eggs of the host bird lie abandoned at the edge of the nest.

Left: A newly hatched cuckoo has an instinctive urge to remove both eggs and young nestlings from the nest.

numerous caterpillar hairs accumulate on the inner walls of the gizzard which periodically sloughs off and is regurgitated.

Cuckoos and Other Parasitic Species

The only other European cuckoo is the great spotted cuckoo (*Clamator glandarius*), which is larger than the common cuckoo, measuring up to 42 centimetres ($16\frac{1}{2}$ inches) with a distinct crest and long grey tail. It is resident during the spring and summer months in Spain and along the Mediterranean coast of France where it inhabits generally open country with scattered groups of pine trees. This species parasitizes birds larger than itself, mainly members of the crow family including the hooded crow, magpie and the azure-winged magpie which is in Europe confined to the Iberian peninsula. Its approach as a brood parasite differs markedly from that of the common cuckoo. Whereas the flight of that bird is thought to deter other birds, that of the cuckoo closely resembles its hosts, especially magpies. Moreover, although the cuckoo's egg does resemble that of the host bird, on hatching it does not forcibly remove the other eggs and young but is brought up side by side with the young crows or magpies. There is often more than one great spotted cuckoo chick reared in the same nest. To compete effectively for food the head parts of the young cuckoos are dark like the young of the host bird whereas the adult cuckoo has a light grey head. The related black-and-white cuckoo (*Clamator jacobius*) of tropical Africa shows a colour reversal – here the young have light grey heads to match those of the young shrikes or bulbuls whereas the adult cuckoo has by contrast a dark head.

Apart from those mentioned there are over forty other cuckoo species that are brood parasites on other birds as well as double that number that brood and feed the young themselves. One of these is the well-known cartoon character 'the roadrunner' (*Geococcyx californicus*), a ground cuckoo found in the American south-west.

Parasitic behaviour also exists in other quite unrelated groups of birds including the honey-guides which parasitize hole-nesting barbets and starlings and the wydah birds which parasitize finches. There is even a South American duck (*Heteronetta atricapilla*) which lays its eggs in the nests of other water birds, notably coots.

The Seal
Sleek socialite of the ocean

Breeding Cycle

Early in June, on the shores of the American Pribilov and Soviet Komandorskiye Islands, two small groups in the Bering Sea, large, heavy animals begin to haul themselves laboriously from the water. The islands are in the same latitude as Britain, but they are remote, bleak and inhospitable. The animals require no food or water from them, however, only privacy. At sea their enemies, the killer whales, take many of their number. In winter the polar bears range widely over the frozen ice, and are dangerous, but by June the islands are surrounded by sea the bears prefer not to cross. Their only land-based enemy is man, and their history has not prepared them for defence against his rifles and clubs.

They feel safe and as they establish themselves on land a curious ritual begins. Although there may be many of them, they do not crowd together. Rather each individual seeks to establish a territory, a stretch of beach from which neighbours are excluded. There is much bellowing, much rearing up aggressively, and more than a little hard fighting that leaves flesh torn and bleeding before the territories are defined, the invisible boundaries drawn, and the animals can begin to relax. The biggest and strongest of them win the best territories – stretches of open sand. The weaker members of the colony must make do with sites further inland or closer to the water, sometimes below the high-tide mark so that for some of the time the animals lie with only their heads above the

Below: A bull Hooker's sea lion (*Phocarctos hookeri*) guards over his harem.

water. Nevertheless, each of them is now a 'beachmaster' and he waits.

In the middle of June more animals begin to arrive, and this time there are far more of them. They, too, haul themselves from the water and as they do so each newcomer enters the harem of one or other of the beach-masters.

These are northern fur seals (*Callorhinus ursinus*), and it is the time of year when they must come together to mate. The two colonies, in the two groups of islands, remain distinct, as though the political frontier separating the islands can be crossed no more easily by a seal than by a human.

Shore Leave

The beachmasters are the males and the second group to arrive are the heavily pregnant females. The gestation period with them, as with all seals, is about seven months, but soon after the embryo has begun to develop further growth ceases, and resumes later, so that the seven-month pregnancy is stretched out to occupy about eleven months. It means that virtually all adult female seals found at sea are pregnant.

The old bulls do not have things entirely their own way, however. While it is they who control the beach, the females are perfectly free to choose the site for the birth of their pups. They go to the stretch of beach that seems to them most suitable. This brings them within the territory of one or other beachmaster, but in order to be a successful beachmaster a male must establish himself on a part of the beach that will attract females. Once the females are there the bulls will try to herd them, but this is only partly effective and the cows remain quite free to come and go and to move from one territory to another – at least until the birth of the pups ties them to a particular location.

There is a perfectly good reason for this. Adapted though they are to life at sea, their adaptation has not proceeded so far that they can give birth to their young in the water – as the whales and porpoises can. The pups must be born on land. However, their adaptation to swimming has led to the limbs of seals becoming flippers, and the hind limbs of the true (phocid) seals cannot be turned forward so that on land the animal must drag itself along, slowly, clumsily and, more to the point, quite defencelessly. In the water a seal will fight hard, and often win, or escape by outswimming its adversary. On land the only way it can respond to a threat from anything more formidable than another seal is by plunging into the sea.

There is a clear advantage, then, in reducing to the minimum the amount of time the seals must spend ashore, and one way to achieve this is for the pups to be born and for the females to become pregnant again during the same enforced spell on land. The arrangement brings the further advantage of facilitating mating. Although most seals are fairly gregarious, mating while at sea depends to some extent on chance encounters between males and receptive females. On shore the males and females – all of whom are receptive at the same time – are close together.

After a day or two the pups are born, and when they are about a week old, their mothers mate again. They mate only with their own beachmaster, and a beachmaster may have as many as fifty females in his harem. Of course, roughly half of all the pups born are male, so that the ratio of one dominant male to fifty females leaves the great majority of males without mates and without hope of mating. They, too, come ashore, but remain some distance from the breeding colony: some of them to fall victim to human hunters who seek their fur. The more courageous spirits among them will try to obtain females, but in the world of the seals this requires them to provide a territory on the beach in which the young can be born and raised and from which rival males can be driven, and the time to establish such territories is before the arrival of the females. It is then that the old bulls may be challenged by younger animals – and each year some of them will be defeated and their stretches of beach will acquire new masters. Once the business of reproduction has commenced it is too late.

So rigidly are the boundaries drawn and observed that it is possible for a human to pass unmolested along them, since in order to attack, the males on either side would have to invade the territory of their neighbours.

Precocious Pups

As soon as each pup is born – and a seal cow

Below: Subject of much controversy – a young harp seal (*Phoca groenlandica*). Each spring thousands of these pups are killed by Canadian hunters despite growing outcries from conservationists.

nutrients they need until they begin feeding. Cautiously at first, but then more confidently, they enter the water alongside the adults, and before long they are full, but juvenile, members of the colony. From that moment on, it is each animal for itself. If it survives the first year or two of life, during which mortality from predators and accidents is high, the young seal may expect to live for up to forty years. As fur seals have been remorselessly hunted for their pelts, attempts are now being made to re-introduce them to islands they once inhabited.

The northern fur seals leave their rookery as soon as the pups are weaned and the adult females are pregnant again, and begin a long migration. The Russian seals move south-west, sometimes as far as Japan, while the American seals move south-east, to the coasts of California. There they feed through the winter until it is time to begin the journey north again, back to the breeding grounds. The round trip, for both populations, is about 10,000 kilometres (6200 miles), and they spend eight months at sea.

Seal Species

Not all seals spend so much of their time far out at sea, and most do not migrate over the long distance travelled by the northern fur seals. Most of them are gregarious, though, the only important exception being the Ross seal (*Ommatophoca rossi*), which congregates only for breeding and spends the winter quite alone in the Antarctic darkness. A seal rookery, of any species, may contain anything from a few individuals to a million or more animals. The Pribilov rookery usually accommodates about a million and a half seals.

Altogether there are about eighteen species of true seals, thirteen species of sea lions and fur seals, and one species of walrus. The true or 'phocid' seals have a thick layer of blubber (fat) beneath the skin, which provides insulation, have no visible external ears, and their hind limbs cannot be turned forward to assist them to move on land. The sea lions and fur seals can turn their hind limbs forward, have small external ears, and many of them have a fur coat and lack blubber. The walrus in some senses is intermediate, having no external ears, but being able to turn its hind limbs forward. Some seals, such as the common or harbour seal (*Phoca vitulina*) never move far from land, and one species, the 1-metre (40-inch) Ions Lake Baikal seal (*Pusa sibirica*) lives in fresh-water.

Above: A dominant male elephant seal holds one of his females before mating.

gives birth to a single pup each year – its mother smells it intensely, continuing to do so for about five minutes. Probably the close contact immediately after birth forges a bond between mother and pup, but the repeated sniffing impresses on the female the scent of her own offspring. Before very long the pup will be feeding, and wandering, and it is prepared to suckle from any female it finds. Like all young animals, frequently it feels lost and cries for its mother. Its cries elicit a response from all the females who hear it, and each of the females it meets will sniff it, but only its mother will permit it to suckle. A lost pup that fails to find its mother may die.

Seal milk is made half from fat, so it is highly nutritious, and the pup grows rapidly. While she is nursing, the mother never leaves her own place on the beach, and she takes no food. After about three weeks the pup begins to lose the soft, and in most species white, fur with which it was born and to acquire its adult coat. It is ready now to be weaned, a rather brutal process whereby the young are aban-doned by their mothers, who return to the sea to begin feeding again. The pups must now fend for themselves.

They are not taught how to do this, but nature has provided them with some assistance. Their diet of milk has given them a thick layer of body fat which will supply them with the

113

The Mole

Hermit of the subsoil

If there is one thing a mole hates it is another mole. The mole is perhaps the most solitary of animals. It must meet in order to mate, of course – though how it does so without starting a fight is something of a mystery – but apart from that it lives entirely alone. It prefers it that way. When two moles meet a violent argument ensues and so, like all solitary animals, quarrels are usually avoided by reducing encounters to a minimum. Among animals that live underground, in tunnels, this is not difficult, since encounters are most likely to result from tunnelling into a neighbouring system. If this should happen, the intruder would block up the hole at once.

Talpa europaea, the common mole, can be found – or at least evidence of it can be found – everywhere in Europe except for Ireland, which it has never invaded. Its hills are a familiar, and sometimes exasperating, sight in fields, beside paths, on lawns, and sometimes on tennis courts, where moles have been known to tunnel along the whitewash lines, probably because the lime in the whitewash alters the chemistry of the soil just a little in their favour. It is quite common for molehills to appear in precisely the same place year after year. This suggests not that moles always follow the same track when digging, but that they spend a good deal of their time repairing old tunnels and that a mole does not tunnel at all if it has a tunnel system adequate for its needs.

Right: The common mole (*Talpa europaea*) constructs a two-tiered burrow system. The deeper level is its living quarters, while the tunnel nearer the surface consists of an extensive network of feeding burrows.

There is plenty of evidence to support this. As soon as they are weaned, young moles must leave home and seek their fortunes. It happens sometimes, inevitably, that an individual inherits the home of one of its parents. Provided this home has been maintained properly it will not tunnel further. Probably there are many moles that have never dug a new tunnel in their lives.

They are prodigious tunnellers when the need arises, and their bodies are perfectly adapted for the task. Their fur can lie toward the head or tail equally well, so they cannot be 'rubbed up the wrong way', and their bodies are almost perfectly cylindrical, but there is much more to it than this. A mole at rest holds its hind legs in the 'knee bend' position used by most small mammals. However, it is able to rotate them outwards so that they stick out almost at right angles to its body, and in this position they brace the tunneller against the sides of the tunnel. At the same time it can rotate its forelimbs so that its paws face outwards and it can flex its paws and stretch its toes apart. Usually it digs using one paw at a time, but in very soft soil it will use both together, proceeding through the soil as though it were swimming a kind of breaststroke.

Life Underground

Living underground as it does, day and night have no special significance for it, but it does have a daily rhythm, based on an eight-hour cycle. During a typical eight hours a mole may rest for about three and a half hours and while it is awake actual digging occupies between one-third and two-thirds of its time.

It has two forms of rest. It takes naps, up to twenty minutes long, wherever it happens to be in its tunnels, but its equivalent to a night-time sleep, lasting from two to nearly five hours, is spent in a special nesting chamber.

When it is not asleep or digging it is patrolling its tunnels. You can tell how active a mole is by finding its tunnels – sometimes they can be traced from the surface by connecting molehills – and making a hole in one with a small stick, then timing the mole to see how

Right: The female mole lines the breeding chamber with dried grass and leaves. Four to six young are born some six weeks after mating.

long it is before the hole is repaired. The tunnels are not simply its home, they are the traps with which it obtains its food. Small soil animals that fall into the tunnels while burrowing are collected by the attentive mole. Most of them are eaten at once, but moles also store their favourite food – earthworms – by biting off the final few segments. This does not kill the worm, but for some reason it prevents it from burrowing. Provided sufficient food falls into its tunnels the mole has no reason to extend them, and it seems likely that moles dig most extensively in soils that contain little food. In the course of a year a mole may consume between 18 and 36 kilograms (40–80 pounds) of invertebrates, probably eating up to 96 grams ($3\frac{1}{2}$ ounces) of food a day – which is about the weight of an adult mole. It does not eat slugs or millipedes.

Although they are subterranean animals, moles swim well when the need arises – as it does when the fields in which they live are flooded – and they do spend some of their time on the surface. In very dry weather, when tunnelling is difficult, they spend much of their time above ground. They surface cautiously. First a snout appears, and that is all that appears until its owner is satisfied that all is well. Then the rest of the mole emerges. It is not blind, but its eyes are small and probably it

cannot see fast-moving objects. It can hear low-pitched sounds and probably high-pitched ones, too, for moles are often vocal themselves. They make a twittering noise when they find food, and a raucous cry when frightened.

The sense organ on which they depend, however, is located in the snout. It is called the 'Eimer's organ' and it is possibly the most complex apparatus in the entire animal kingdom. Its full function is not known, but it is believed to be highly sensitive to touch, chemical stimuli, pressure and temperature.

A mole lives for two or three years. Like most small mammals it wears down its teeth by biting on gritty, abrasive food, and if it dies from no other cause eventually a time comes when it cannot eat and so starves. It has been calculated that for this reason it is impossible for a mole to live for as long as five years.

To the naturalist the mole is a fascinating small mammal, but to the farmer it is a serious pest. On pasture a mole's tunnelling disturbs the roots of grasses and clover making them more susceptible to drought and frost. On arable land the damage is even worse as they 'plough' their way through newly sown root crops. Many farmers desperate to rid their land of moles, use poisons which, though effective, also kill birds of prey which feed on the carcases.

The Penguin

High society of the polar seas

Far out in the southern ocean the flock of Adélie penguins (*Psygoscelis adeliae*) spends the winter feeding, hundreds or even thousands of birds together. They keep clear of the ice that covers the winter seas around the shores of Antarctica and in the course of a year they may swim a complete circuit of the continent, returning late in October – in the southern spring – to the point from which they departed months earlier.

Although they departed more or less together, by the time they return the colony has broken up into many flocks. It is the males which arrive first, leaving the females to continue feeding for a few days more. The birds are plump, for they have eaten well and have accumulated a large store of body fat that will sustain them during the period that lies ahead.

Nesting and Breeding

The males bustle ashore and begin to search for the nests they built in previous years. The rookery site may be close to the shore,

but Adélies have been known to travel more than 100 kilometres (62 miles) across the ice to their traditional breeding ground. They cannot breed on the ice itself, because this offers no nesting materials, and Adélies build nests.

The bird that finds its old nest sets about repairing it, using stones or bones as building materials, and driving away males that have no nests who try to take it for themselves. The nests are not completed, for once the nesting sites have been claimed and work has begun the females arrive.

The task now is for the penguins to form pairs. Like all penguins, Adélies mate for life, and the social life around the rookery begins with the search for old mates. It is the females who do most of the searching, for the males must stay to guard their nests. Inevitably, the winter has taken its toll of casualties and not every penguin finds its former mate. In such cases new mates must be chosen.

The unmated male calls loudly – and a penguin rookery is a noisy place – his head pointing upward, and he waves his flippers slowly in an elaborate display. An unattached female, attracted to him, approaches the nest, stops a little way from it, stares hard at the male and then bows. The male returns her stare and her bow, then lies down in his rudimentary nest and makes movements with his feet that suggest hollowing out a space in which eggs might lie. If the female accepts his proposal she will begin to help with the nest building. At this point it is not unusual for the male's former mate to arrive, late but undaunted, and to send the usurper packing.

Such marital fidelity among birds may strike us as charming, or quaint, but it serves a useful purpose in a region where the breeding season is made short by the return of snow and ice that quickly buries nests. Where the partners know one another already many of the courtship rituals may be omitted. The rituals are important, for it is by means of them that mates are chosen for some quality that will favour the survival of the young. Their omission increases the time that is available for

breeding and for rearing the chicks. The more advanced they are by the time they leave the rookery, the better will be their chance of returning to it the following year.

While they are building, the penguin pair must keep an eye on the stones they have collected, for thieving is common. The stone thief will seize his booty and steal away into the crowd, his feathers held close to his body to make himself look smaller, as he tries to become invisible. The victim of the crime will raise his feathers in outrage and set off in pursuit. The pursuit may lead to further territorial infringements and quarrels, but if he finds the felon he will give vent to his feelings in a stream of abuse that may include a few slaps with his flippers. The thief will accept the rebuke patiently, making no attempt to defend himself, and so the precious stone will be returned to its rightful owner.

November is well advanced by the time the penguins have completed their nests and preliminary courtship, and the female has laid her two or three eggs. During the month that has elapsed since they left the sea neither bird has eaten anything. They have lived off the fat stored in their bodies. The lack of food has affected the female more severely than the male, because she has had to devote some of her bodily reserves to the production of the eggs. She will have lost up to one-third of her body weight. Consequently it is the male who lies on his front on top of the eggs and commences the incubation, while his mate returns to the sea. She stays away for about two weeks, feeding voraciously until she is plump again. Her food consists mainly of krill – small shrimp-like animals that are abundant in Antarctic waters. Meanwhile, back at the rookery, the male must guard the eggs against predators; skuas and rival penguins.

Skuas are large seabirds that arrive on the scene soon after the eggs have been laid and patrol tirelessly in search of unattended eggs or chicks. There is no other food for them, but their search yields sufficient to make it worth their while, and penguin losses are high. Penguins also steal eggs from one another, not to eat them, but to incubate them. The fact that a bird has no mate or nest does not weaken its urge to be a parent. The rookery contains many such birds and they spend their time wandering among the nests, ever ready to appropriate the family of another.

The return of the females is marked by formal greetings – by which the pair identify one another – and then it is time for the male to depart to feed. Before he goes he checks the nest and makes any repairs to it that are necessary. Then he leaves for the krill and, like his mate, spends about two weeks feeding. After his return to the nest the birds take it in turn to lie on the eggs, the bird that is not incubating spending its time at sea.

The chicks are born after about thirty-four days and until they lose their downy coating and grow waterproof feathers they must be brooded to keep them warm. They also need to eat a great deal. In its first two weeks a chick will increase its body weight from about 85 grams (3 ounces) to more than 1 kilogram (2.2 pounds) and by that time it will be demanding well over half a kilogram (1.1 pounds) of food at each meal, and several meals a day. Its rapid growth helps it to survive by reducing to a minimum the period that must elapse before it can take care of itself.

Soon it is too big to fit in the nest and leaves to join the other chicks, which spend their time together, huddling for warmth when the weather is cold, and being fed by the adults. As soon as it has feathers it is ready to take to the water. It must learn the correct penguin way to enter the sea, for the sea can contain leopard seals which eat unwary penguins. The technique is for the penguins to gather at the edge of the water, jostling one another until one of them falls in. If the victim is eaten by a seal the others wait a while then resume the jostling until another penguin falls in. As soon as one survives its plunge and the penguins see that the water is safe, all of them enter.

The Extraordinary Emperor
Most of the fifteen species of penguins live in this way, all of them in the southern hemisphere from Antarctica as far north as the tropics, but only the Adélie and Emperor (*Aptenodytes forsteri*) breed on the Antarctic mainland. The Emperor differs from all other penguins because although it uses the same breeding sites as the Adélies, it breeds in winter rather than summer, and incubates its eggs without making any nest, in temperatures that may fall to $-80°C$.

After mating, the female lays one egg. She incubates it by resting it on her feet and covering it with a fold of skin, but at first she lifts the skin frequently to show the egg to

Below: The Adélie penguin is one of the world's southernmost breeding birds, congregating on ice floes and rocky hillsides of the Antarctic continent. Here a group of Adélies wait their turn to leap into the sea.

her mate. Then he takes the egg from her and incubates it and for the first few days the two pass the egg back and forth frequently. Then the female departs to feed and the serious business of incubation begins among the males who are left behind. They huddle together in a circle, shuffling very slowly to the side so that the entire circle rotates – and the icy wind blows equally on all the birds. As a bird at the outside becomes cold it pushes its way toward the centre. So the huddle is moving constantly, and the temperature of the eggs is held at a steady 30°C. Eventually the females return, with their crops full of fish, and take over the incubation from the first male they meet – not necessarily their own mate. When the chicks are born they must be brooded until they are large enough to form groups similar to those formed by the Adélie chicks. In spring the Emperors migrate to South America.

The Baboon

Safety in numbers

The ancient Egyptians held it to be sacred. Modern farmers regard it as a pest that causes considerable damage to crops. Travellers are warned that if molested it can be dangerous. The baboon is an animal that cannot be ignored by humans who share its range.

It is fairly large and an adult male is armed with long canine teeth which it will not hesitate to use if threatened. It is not the strength or armament of the baboon that makes it a formidable opponent or rival, however, but its social organization and its intelligence. The baboon never lives alone and the sight of one is a sure indication that a full troop is close by, within call, and ready to react at once to anything that approaches. A young animal, or a female, enjoys the protection of the biggest and strongest members of the troop. Apart from humans armed with rifles, the only real enemy the baboon has is the leopard.

Today baboons range over the grassland and dry, rocky places from the southern edge of the Sahara Desert to the tip of southern Africa, and one species – the hamadryad which so impressed the Egyptians – is found in the south-eastern tip of Arabia (Yemen) as well as in Ethiopia and Somalia. Probably the ancestors of baboons once lived in the forests, where such close relatives as the drills and mandrills live today, and as the ancient forests retreated

Right: A group of hamadryad baboons (*Papio hamadryas*). The male hamadryad is the most powerful and aggressive of all species, maintaining strict discipline within his group. Any youngster that strays away is quickly punished with a sharp bite on the neck.

because of a deterioration in the climate, some of them were able to modify their behaviour sufficiently to enable them to survive in more open country. The modification must have been considerable, for without the cover provided by trees animals are much more exposed to the unwelcome attentions of the large carnivores.

Group Organization

To survive in the open a high degree of social organization is needed so that some individuals stand watch while others feed. It is difficult to dig up roots, pick leaves and fruit from shrubs, catch insects or steal eggs from the nests of birds and at the same time to be alert to the tell-tale movements of the grass that betray the presence of a stalking leopard. Yet such advanced organization is possible only for a highly intelligent animal. Many of the large grazing animals of the plains can rely on the speed at which they run to escape from

danger, and each adult individual is large enough to be able to see danger from a distance just by raising its head, so that in a herd at any moment there are bound to be a few animals looking around while they chew. About the size of a large dog, the baboon is too small of stature to be able to see far without climbing on to a rock or a fallen tree, and it cannot run fast enough to escape an attack unless it has ample warning. The animal that keeps watch cannot feed at the same time and so, for the good of the troop, particular individuals must wait to feed. Such selflessness is one mark of high intelligence, but there are many others.

The size of a baboon troop varies from place to place. In Namibia, for example, there is an average of twenty-seven animals of all ages and of both sexes in a troop. In Zimbabwe the number is about forty-six, and in Kenya there may be as many as eighty.

The troop lives within a range – the area within which it obtains all of its food. The core area, where the baboons sleep, contains a clump of tall trees, and the animals climb into the trees at night. During the day they forage. In dry weather they will dig down 40 centimetres (16 inches) into the soil to find roots, tubers and bulbs – and they have a particular liking for groundnuts (peanuts). They prefer leaves and fruit, however. It is their predominantly vegetarian habits that make a full baboon troop a menace should it invade a farm. Were the opinion of the baboons sought, of course, they might well argue that it was the human farmer who had invaded their range and not the other way round! For the farmer, though, the problem is made worse by the fact that like many Old World monkeys that feed mainly on the ground, baboons have large cheek pouches in which they can carry back to their core area any food they are unable to eat on the spot. They can strip an orchard bare, and they are liable to attack the farmer who tries to prevent them from ruining him.

They will eat insects, but only when insects are plentiful and vegetable food is scarce, and they will take eggs. Occasionally they will join together in attacking an animal such as a hare, or even a gazelle, which they will tear to pieces with their hands.

The Dominant Male

In all baboon troops, except for those of the hamadryad, a number of adult males live side

individual who is senior to them. The society is distinctly chauvinist, for the females enjoy no rank whatever, even among themselves – although having a small infant to care for entitles a female to some minor preferential treatment. All females are subordinate to all adult males, a young male being able to dominate females from the time he is four years old.

The fact that they are able to dominate females socially does not entitle the junior males to mate with them if they are mature and on heat. They may mate freely provided pregnancy does not result, but four-fifths of the young are fathered by the dominant male. Such fruitful matings do not lead to any permanent relationship between the male and his females: it is the fact of the female being on heat that attracts the male, and the association lasts only while she remains on heat.

The dominant male is very aggressive, to members of his own troop as well as to outsiders, but a large part of his aggression is intended to protect the troop, and mothers and infants come to him when anything frightens them. The young males, too, spend a good deal of their free time fighting. It is only by overthrowing the baboons ahead of them in the hierarchy that they can gain promotion and only by defeating their peers that they can come to dominate them.

On the Move

When the troop moves about its range the members arrange themselves in a formation that provides the greatest possible security. The infants stay right at the centre of the group, surrounded by their mothers and other adult females. The dominant male, and other senior males, stay at the centre close to the females and young. The remaining adult males position themselves around this group, in front, to the rear, and to both sides, but the juveniles – the immature males and females – remain on the outside of the troop, to either flank. These young animals are agile and should the troop be attacked they can make a run for it as there is nothing in their way. The central group is protected by the adult males, and the adult females and infants are protected immediately by the dominant males. There is no easy way to attack a troop of baboons on the move!

The Hamadryads

Among the hamadryad baboons the social

by side, together with all the females and the young. There is a social hierarchy among the males, so that some dominate others and all of them are subordinate to a single male who leads the troop. The social organization is not entirely rigid, for now and then two or more males may form an alliance to overthrow an

groups become permanent, with the females always following their respective males. Should the male be lost or defeated in combat, the females will begin to follow another male, sometimes selecting one of the bachelors for this role. Among the hamadryads it is herding that matters. A group leader will allow other males to mate with his females, but he will not allow them to be herded.

Obviously, such elaborate social organization requires a great deal of communication. Some of this is visual, but most is vocal. Various bodily postures and positions of the tail have definite significance, but in addition there is a wide range of sounds, from a bark that warns of danger, to grunts, squeals and chatterings, each of which has a meaning.

Species and Range

Despite its bark, and despite its dog-like face, the baboon is not related to the dog. Nor is it an ape, despite its great intelligence and the readiness with which it learns new behaviour and profits from experience. It is a monkey. Strictly speaking, the word baboon is applied only to five species, all belonging to the genus *Papio,* each species being confined to particular parts of Africa. The chacma baboon (*P. ursinus*) lives in southern Africa, the yellow baboon (*P. cynocephalus*) in Central and East Africa, the olive baboon (*P. anubis*) right across the southern edge of the Sahara, with the Guinea baboon (*P. papio*) in a small enclave on the West African coast and the hamadryad baboon (*P. hamadryas*) to the east. More loosely, the word is used of some other monkeys from Africa – most especially the gelada, which lives in a similar habitat and looks rather like a true baboon – and one from Indonesia.

Far left top: Mutual grooming plays a large part in the life of the gelada baboon (*Theropithecus gelada*).

Far left bottom: The cohesive nature of a troop is shown by these female baboons. The baby's pink ears and face stimulate the mother's maternal instincts.

Above: Excursions for food are strictly organized manoeuvres. The strongest young male baboons act as scouts and form the vanguard while other young males patrol the flanks and rear. The dominant males form an inner ring of security at the centre of the troop.

organization is slightly different. The hamadryad troop sleeps all together in the core area, but during the day it divides into several groups, each group consisting of an adult male and up to six females. In addition there is a group composed entirely of bachelor males. When they are three or four years old, male hamadryads begin to herd females. They do this by taking juvenile females from their mothers – just one or two at first, but after a couple of years a young male herds more and begins to adopt infants of both sexes. In this way the animals become partners before any of them is old enough to mate. A male hamadryad meeting a strange female may try to herd her, but he will not try to mate with her. The

The Solitary Wasp

Lone hunter of the insect world

Wasps can be divided by their habits into social and solitary types. The former are the familiar wasps and hornets which live in colonies, build elaborately constructed nests, and often appear very numerous especially at the end of summer. These insects are, however, only numerous in terms of the number of individuals; a far greater number of species are not social but lead solitary lives.

While the common wasp supplies its larvae with a constant supply of masticated food – other insects which are seized and mashed to a pulp by the wasp's strong mandibles – the solitary or hunting wasps, in contrast, paralyse insects so that their tissues retain their freshness until the eggs hatch out and they are consumed by the developing larvae.

Each genus of solitary wasp tends to specialize in a particular prey: *Philanthus* hunts bees, *Ammophila* caterpillars and *Sphex* tackles large grasshoppers and crickets. The number of hunting wasps reach their greatest development in the warm dry climates of the northern hemisphere particularly round the Mediterranean and Central America.

British and European Species

The most comprehensive account of hunting wasp behaviour was made a hundred years ago by the French naturalist Jean Henri Fabre who in his later years lived in Provence where the light sandy soil, and abundance of prey species provided ideal conditions for his observations. Many of his studies were conducted in his large garden – actually a piece of waste land – which he stocked with a wide range of wild plants to attract different types of insects.

There are fewer species in colder climates but usually there are a number of types that are large enough and common enough to study in the wild. The potter wasp (*Eumenes coarctatus*) is the only one of its genus in Britain and is restricted to the heaths of southern England. The female builds an elegant flask-shaped nest reminiscent of a Grecian urn with a short neck. The pot is made of sand and small pebbles and often glitters with small pieces of quartz and

shell. It is generally constructed on the end of a sprig of heather and the wasp may build several within close vincinity. When the structure is completed, she goes off to hunt, collecting only small, smooth green caterpillars which can easily be pushed through the narrow opening. About a dozen or so are stuffed into each pot. After this, because the entrance to the pot is too narrow for her to enter, she inserts her abdomen, and with ovipositor extended, she lays a single egg which hangs by a silken thread just above the provision of paralysed caterpillars. It appears that unlike many wasps, *Eumenes* is not an expert in subduing the caterpillars, and so, suspended above its future meal, the egg is protected from being squashed by a caterpillar's occasional writhing spasms. On hatching, the young larva tackles the nearest caterpillar, then, released from its thread, it eats its way through the whole stock before pupating. When the adult wasp emerges, it chews its way out of its cell, mates and re-enacts the life cycle.

All wasps of the genus *Pompilus* hunt spiders. *Pompilus naticus* prey mainly on wolf spiders, while the British *Anoplius fucus* is said to hunt mainly web-building spiders. Certain American wasps even tackle large tarantulas which they haul over the ground until they find a suitable spot to build a burrow. Some pompilids can walk about a spider's web without getting snared, others angrily buzz around the web threatening the spider until it is forced to drop to the ground where it is quickly dispatched. After the prey is paralysed, a female will generally hide it on a blade of grass to prevent other insects, especially ants, from carting it off. She then digs a burrow and between bouts of strenuous activity checks to see that her victim is still in place and to memorize the position of prey and burrow will often fly around the area repeatedly. When the burrow is dug, she drags the spider down, lays eggs on it and carefully seals the entrance by tamping down the sand with her feet and abdomen.

The wasp *Bembix* is rather different in its catchment of food. It is known as a progressive

provider in that the young larvae are given a constant supply of fresh food in the form of flies. Perhaps this method has arisen because of the insubstantial nature of a fly's body which would soon dry out if left for any period of time. She first chooses a small fly on which to lay her eggs and as the larvae hatch out and their appetites increase, supplies her developing young with larger food items including gadflies which includes the large blood-sucking horseflies.

Fabre gives an amusing account of how, when stationed by the roadside in a wood, 'wasp watching' under the shade of a large umbrella, he experienced the hunting techniques of the female *Bembix*:

On certain afternoons in the dog-days of July and August I used the shelter of a large umbrella. I was not the only one to profit from the shade; I was generally surrounded by numerous companions. Gadflies of various species would take refuge under the silken dome and sit peacefully on every part of the tightly stretched cover. . . To while away the hours when I had nothing to do, it amused me to watch their great golden eyes, which shone like carbuncles under my canopy.

One day, bang! the tight cover resounded like the skin of a drum. Perhaps an oak had dropped an acorn on the umbrella.

Presently one after the other, bang, bang, bang!... I looked up and the mystery was explained. The *Bembix* of the neighbourhood was impudently penetrating my shelter to seize flies on the ceiling. I had only to sit and look. Every moment a *Bembix* would enter, swift as lightning, and dart up to the silken ceiling, which would resound with a sharp thud. The struggle did not last long, and the wasp would soon retire with a victim between her legs. At this sudden interruption which was slaughtering them one after the other, the dull herd of gadflies drew back a little all round, but did not quit the treacherous shelter. It was hot outside! Why get excited?

Britain does not have a great number of hunting wasps but the behaviour of several species can be observed, especially on light sandy soils and heaths in southern England; old quarries are also good places to observe them. Because the Mediterranean region is so rich in species, if on holiday there it is worthwhile getting away from the beach for half an hour into the garigue to watch these fascinating insects.

Above left: Potter wasp (*Eumenes coarctatus*) stocks its egg cell.

Top : A species of *Bembix* – a genus of wasps that are specialized fly killers. Bottom: Sand wasp (*Ammophila pubescens*) carrying a paralysed caterpillar.

SOCIABLE AND SOLITARY ANIMALS

Name	Habitat	Social structure
Common seal *Phoca vitulina*	Along coasts in the northern Pacific and Atlantic Oceans. Come ashore in groups usually on mud and sand banks, sometimes on rocks.	Sociable and playful, common seals do not hold territories and only rarely fight. Colonies in Britain may contain 400 seals, perhaps half of breeding age.
Common mole *Talpa europaea*	Moles dig out and live in burrows in any suitable soil, under woodland or more open country if it is not waterlogged. Home range of about 400 m².	Aggressive and solitary, captive moles always fight, and if one moves into another's burrow, it will be fought off. There are some communal burrows, for example where narrow earth banks cross ditches.
Baboon *Papio anubis*	Roam in groups of 30–50 individuals in open country or scrub and woodland. Home range of a troop 5–40 km² depending on the availability of food.	Each troop includes several adult males and breeding females, a group of young animals and some infants. If they live where food is abundant, much time is spent playing, grooming each other and dozing, but in dry areas most time is spent food gathering.
Tiger *Panthera tigris*	Tigers live in thick forest, woodland or long grass in many parts of Asia.	Tigers are almost never seen together except when mating. They do not have an exclusive territory but several may wander in the same area. They mark trails with scent to communicate their movements. If two meet they may growl, but usually withdraw from the encounter by turning away.
Wolf *Canis lupus*	Wide-ranging and adaptable animals, wolves live in much of North America though their range in northern Europe has been much reduced in historical times. Packs of 2–20 individuals occupy territories varying from 300–1200 km² depending on the abundance of prey.	A wolf pack is essentially a family group. There is a well-marked order of dominance, with senior males leading the cooperative hunts. They communicate by howling (which keeps the pack together) and by complex visual signals as well as scent.
Lammergeier *Gypaetus barbatus*	Live in moutainous regions, with several eyries. Have a large home range, about 300 km² for a pair in the Pyrenees.	Largely solitary, especially while hunting. The largest and rarest European vulture, it feeds on freshly dead meat and bones of mammals and reptiles.
Adélie penguin – *Pygascelis adeliae*	In winter swim in groups at the edge of the Antarctic pack ice, and nest in large rookeries on land.	Social at sea and on land. Penguins call to keep together at sea. When in the rookeries there is much social interaction to maintain the territory.
Cuckoo *Cuculus canorus*	Breeds in northern Europe and Asia, where it lives in a variety of habitats, open woodland, thickets and moors. Winters in central and southern Africa, India and S. Asia in thorn bushes.	Cuckoos are solitary birds, and the well-known call of the male during the breeding season helps to keep the population spaced out.
Starling *Sternus vulgaris*	Abundant in northern Europe and Russia, introduced into North America, starlings live almost anywhere in lowland habitats, town or country.	Feed in close flocks on the ground, flocking especially closely if a bird of prey such as a sparrow hawk flies over. The largest flocks in the autumn countryside may contain over a million birds and will be made up of residents and migrants.
Solitary wasp *Sphex lucae*	This North American solitary wasp lives in open woodland and scrub.	Solitary, territorial insects, which range quite widely for food but return to familiar landmarks to find the burrows they dig for nesting and in some forms for sleeping.
Honey bee *Apis mellifera*	The Western honey bee lives in most kinds of country in Europe and North America, nesting in hollow trees or rock crevices in the wild.	Highly specialized social insects with three castes of individuals: the solitary queen; the workers, sterile females which feed larvae, build and clean the nest, collect food and act as guards; and drones, short-lived males. Very complex 'language' of dance and chemical signals as well as touch.

Reproductive behaviour	Number of young
Common seals play especially boisterously in the water in the autumn mating season. It is thought that the same pair remains together at least for one season. They pair at sea, and the young stay closely with the mother for 3-4 weeks, then spend much time playing with other pups.	Usually 1 pup is born to each seal cow.
At the mating season, male moles move outside their usual territory in search of a mate. This is the only period when a female will not attack an intruder at once. The pair stay together for perhaps an hour. The young stay with the mother for 6-7 weeks, during which time they do not fight.	A female mole usually gives birth to 4 young.
Female baboons first mate at 3½-4 years old, the males age 4-6. They mate with males of their own group and similar ranking level, staying together for about 2 days. Young are born in September-February and are fed by the mother for 5-8 months.	The female baboon normally gives birth to 1 young.
A female tiger calls when in season to attract a temporary mate. The courtship is noisy and the male will fight and drive off rivals. When the young are born, the other avoids all other tigers including the father. Cubs stay together and with their mother for 18-22 months.	Cubs are born in litters of 2-6, with 3 being commonly found.
In the summer breeding season, wolves stay in a smaller territory. The dominant female is usually the only one that mates. In a big pack mothers may mate too, and the pack will then probably split. Males and females of all the pack will feed and care for the young when weaned.	Between 4-14 young can be born to one mother.
Lammergeiers mate for life. The indulge in spectacular courtship flights, pursuing each other and falling nearly to the ground with interlocked claws. The nest site is changed each year.	Only 1 fledging is usually reared, and it remains in the nest for 100-110 days.
A pair of Adélie penguins generally remains together for life, returning to the same breeding site. Fledglings congregate in large crèche groups for protection and warmth.	Usually 2 young are hatched.
After mating, the female lays a single egg in the nest of a fostering species and moves on to other nests until she has laid 12 or more. No further care is given to the young and the adults migrate south singly in July leaving the young to follow on their own a month later.	Approximately 12 young, placed singly in the nests of host birds.
Starlings may breed in colonies or single pairs, but do not show marked flocking behaviour in spring. The male sings and flutters his wings to attract a female, which will take the initiative in forming the temporary pair.	The female lays 4-7 eggs which fledge in 20-22 days after hatching.
This solitary wasp pairs briefly in a series of mating flights. The female then lays her eggs in a series of prepared burrows, each with a cell for a single egg provided with a food supply for the growing larva: 3 long-horned grasshoppers she has trapped and carried to the nest.	Varies from species to species. Only 1 egg is laid per nest, the overall number depending on the availability of food for the larvae.
Honey bee queens mate once in their lives in a nuptial flight pairing with several drones. Drones develop from unfertilised eggs, workers and queens from eggs fertilised by the stored-up sperm. Colonies multiply by swarming and splitting with a new young queen.	Up to 40,000 eggs are laid each summer for several seasons.

THE GAUDY
— AND —
THE CAMOUFLAGED

A herd of Grant's zebra *(Equus quagga granti)*.
In thick vegetation the stripes have a disruptive effect,
aiding concealment. In the open the dazzling
patterns may confuse predators chasing
the tightly-packed herd.

The Frigate Bird
Gaudy pirate of the tropics

Far right: The magnificent male frigate bird (*Fregata magnificens*) is spectacular – during the breeding season a patch of naked skin on the underside of the neck turns bright red and is inflated to an enormous size.

Below: Frigate birds are masters of flight. They hang motionless in the air on broad outstretched wings, then will suddenly turn, pursue, and overtake another seabird, forcing it to regurgitate its catch of food.

Frigate birds are the pirates of the tropical seas, using their speed and agility in the air to wrest hard-earned food from other seabirds. With their enormous wings and long forked tails they appear like gigantic swifts cavorting over the ocean surface or swooping down from great heights on their victims.

Their main characteristic feature, which the males of all five species possess, is a throat sac which during the breeding season changes from its normal orange colour to bright red and is inflated like a balloon completely dwarfing the head of the bird. The sac is used primarily to attract females flying overhead and also to warn other males against intruding on the breeding territory. While squatting on the ground with his scarlet sac thrust skywards, the male spreads his sleek black iridescent wings, shakes his body and rattles his bill. The throat sac, apart from its brilliant colour, is thought to act as a resonating chamber to further advertise the male's presence. Once a female is attracted, the pair copulate on the ground and proceed to collect twigs for the nest. These are gathered in flight, either snapped off trees or picked off the ground, but also, characteristically, stolen from the nest of another seabird.

A single large white egg is laid, and once the young hatches out it is guarded by either parent from the attacks of neighbouring frigate birds, who are by no means above a little cannibalism. Although the young are fully fledged at five months, the juvenile birds remain dependent on their parents for a further six months. In the meantime they test their wings and during these preliminary flights they often fly in small bands practising their manoeuvres by trying to catch a piece of seaweed or a feather in mid-air.

Adult frigate birds are true masters of flight. They can soar effortlessly, taking advantage of rising air currents to gain height, and are able to remain motionless on the wing for hours at a time. Like skuas, frigate birds feed most profitably on the catches of other seabirds including boobies (tropical relatives of gannets), pelicans, cormorants and gulls. Once they spot a bird with a full crop of food returning to its nest they pursue it remorselessly pecking at its tail and wings until, worn out by the chase, the bird regurgitates its catch which is quickly swallowed by the attendant frigate bird. When hunting for itself a frigate bird uses its long hooked bill in a snatch and grab technique to take young seabirds that are momentarily unguarded by their parents. A frigate-bird colony is often conveniently situated near a colony of other birds which is periodically raided. Frigate birds also hunt out at sea where flying fish appear to be a favourite food. These fish are taken as they 'fly' out of the water when they are flushed above the surface by marauding shoals of dolphin and large fish such as albacore and tunny. Frigate birds will also take anything edible that is lying or swimming on the surface including carrion, squids and the occasional young turtle – all are flicked out of the water causing barely a ripple on the surface.

On land the grace these birds display in the air is replaced by clumsiness. Their legs are tiny and incapable of walking, so they can only perch on branches or squat on the ground with their legs and feet folded up awkwardly beneath them.

The Chameleon

Master of disguise

Far right: The common chameleon (*Chamaeleo chamaeleon*) is found in countries bordering the Mediterranean.

Below: A Fischer's chameleon (*Chamaeleo fischeri*) from Tanzania blends perfectly against the bark of a tree. This species will hunt lizards smaller than itself.

Colour change in chameleons is not simply a matter of blending in with the surroundings, but is a much more complicated affair involving light intensity, temperature and behaviour. From experiments conducted on laboratory animals two conflicting sets of environmental factors appear to operate. In complete darkness a chameleon becomes pale, almost white, and as the light intensity is increased the skin gradually darkens. At the same time a chameleon will become light-coloured on a light background and will darken on a dark background. Just how a chameleon knows which set of changes is appropriate at any one time remains a mystery, although we know that the skin is sensitive to light and probably also to reflected light from the surrounding vegetation. Apart from natural camouflage and defensive coloration, male chameleons darken prior to mating while females about to lay eggs present a dazzling show of colours.

The actual darkening and lightening of the skin is due to movements of pigment granules inside melanophore cells. When the pigment is dispersed along the fine branches of the cells the animal becomes darker; when the granules are concentrated within the body of the cells the animal takes on a lighter colour.

The range of colour exhibited by chameleons usually only encompasses greens, yellows and browns, but even within this narrow range the change can be amazingly quick and startling. The European chameleon (*Chamaeleo chamaeleon*) which is found from the south of Spain across North Africa is normally a dull-coloured greenish-brown but can instantly change both its coloration and body size. A European biologist relates his encounter with one of these chameleons in North Africa:

> About twenty centimetres in front of my feet was a grass-green chameleon rocking back and forth in an imposing threat display. When the chameleon was picked up it immediately became a very pale cream with brown and black stripes and puffed itself up to half its size again.

The common chameleon of tropical Africa shows an even greater range of colour patterning. This aggressive little reptile when excited

becomes flushed with an overall dark olive-green interspersed with yellow and white spots. When one of these chameleons meets its chief enemy, the boomslang snake, it erects its head shield, hisses loudly and its body becomes stippled with light and dark markings. Apparently a mature boomslang is not deterred by this startling display, but it often serves a protective function in disconcerting other predators.

Feeding and Breeding
Apart from its astonishing ability to change colour, other aspects of a chameleon's lifestyle make them a fascinating group of reptiles. Most chameleons of the genus *Chamaeleo* live in bushes and shrubs where they hunt insect prey. In an adaptation to this arboreal life the toes of the feet are arranged in two groups (two in one and three in another) which grip the branches like tongs. The tail gives added support coiling tightly around a branch like a watchspring. Movement is slow and deliberate – one forefoot is lifted and moved forward to stop just short of clasping the branch. The chameleon then rocks rhythmically backward and forward, grasps the twig with its raised foot then repeats the

procedure using the opposing hind foot. Whilst maintaining slow and cautious progress along a branch the chameleon's eyes, truly remarkable mechanisms, are constantly swivelling around searching for insects. Each eye is enclosed by a scaly cone-shaped lid with a small round opening for the pupil. The two eyes operate independently like guns in twin movable turrets, allowing the chameleon both wide-angled monocular and binocular vision. Each eye can move 180 degrees in the horizontal plane and ninety degrees in the vertical. When an insect is sighted both eyes are focused on it and the front part of the body begins to rock from side to side so that the chameleon can gauge the distance of the prey by viewing it from different angles. It is then that the other remarkable adaptation comes into play, the long sticky tongue.

Before striking the chameleon points its head at the insect and opens its mouth with its tongue slightly protruding. Then it 'fires' and within a fraction of a second the tongue shoots out, traps the insect on the sticky, club-shaped end and brings it back into the mouth. The tongue is operated by two sets of muscles, one set running the length of the tongue and keeping it packed tight on a bone at the back of

Below: Head of a male Jackson's chameleon (*Chamaeleo jacksonii*). The horns are used mainly to intimidate other males but are on occasion thrust at an opponent. The female possesses only a single median horn.

the mouth when not in use. The 'firing' is effected by the relaxation of these long muscles which causes the tongue to shoot forward like a released spring. Added impetus is given by the contraction of the other set of muscles which encircle the tongue. Using this apparatus some chameleons are capable of snaring prey larger than insects; for example the 55-centimetre (22-inch) Oustalet's chameleon (*C. oustaleti*) of Madagascar feeds on small mammals and lizards, and Meller's chameleon (*C. Melleri*) of

proportions – these lizards are generally very small, often less than 5 centimetres (2 inches) long, have a stumpy little tail and live on or near the ground. Although not credited with a great ability for colour change, they also exhibit a talent for disguise. They are often reddish-brown with peaked projections on the head and knobs and spines all over the body, making them appear, to the eyes of a predator, like fallen leaves or pieces of dead wood lying about in the undergrowth.

Left: A chameleon hunts by stealthy approach followed by the split-second release of its sticky tongue.

Below: Jackson's chameleon is confined to the East African highlands. Its colour repertoire includes all shades of green, yellow and brown.

mainland Africa commonly takes young nestling birds.

Of the eighty chameleon species most live in tropical and southern Africa and in the Malagasy Republic (Madagascar). The genus *Chamaeleo* accounts for seventy species, many of which have various distinctive heads shields, horns and foot spurs while some Madagascan chameleons sport peculiar snout appendages. Fischer's chameleon (*C. fischeri*), a lively and aggressive species, uses its large horns in threat displays and in actual combat with other males. Among the group some lay eggs and others give birth to live young. The eggs are laid in the ground where the mother slowly and methodically digs out her breeding burrow. Most of the live bearers come from the montane forests of Central and East Africa and include several brightly coloured species. It should be added that a young chameleon although fully equipped with an extensile tongue often misses its target at the first few attempts and requires practice to become a consistently accurate performer.

There are other chameleons of the genus *Brooksia* which lack the chameleon's typical

Insects Resembling Plants

Now you see them, now you don't

It is no accident that insects which feed on the leaves of trees and other plants tend to be green, while those found amongst leaf litter are usually brown. In the highly competitive insect world camouflage is one of the commonest forms of protection against predators. On closer inspection, one soon realizes that matching body colour to the environment is the simplest form of adaptation in the insect repertoire.

A subtle refinement of the typical green caterpillar is shown in the larva of the European eyed hawk-moth (*Smerinthus ocellata*) which uses the principle of countershading to make itself almost indistinguishable from the sallow leaves on which it feeds. This big green caterpillar is shaded pale above and dark below, but it tends to feed in an upside-down position and the effect of its body shading is to counteract all shadow, transforming its plump shape into a flat two-dimensional one. The illusion that it is a leaf is completed by seven oblique white stripes which closely resemble leaf veins. In South America certain hawk-moth caterpillars show an additional refinement in having two bright white splashes which, in the vertical tropical sun, simulate the light reflected from the shiny leaves of its food plant.

The most convincing green leaf disguise is shown by the Indian leaf insect (*Phyllium bioculatum*), long-horned grasshoppers and mantises. The large flat wings of *Phyllium* are shaped and patterned exactly like leaves but so also are the legs which bear broad green-coloured phalanges. In certain long-horned grasshoppers the large forewings which are leaf shaped very often have irregularities which look as though they have been chewed by other insects. Others contain a marvellous irregular sprinkling of pale spots that closely resemble moulds.

Perhaps even more remarkable are those insects that resemble flowers. African bugs of the genus *Ityraea* are brightly coloured sap feeders that tend to feed gregariously clustering around a terminal shoot of a plant looking innocuously like an unopened lupin flower. But when disturbed the symmetry is

Below: During the day the Central American mantis (*Choeradodis rhomboidea*) rests on the trunks of forest trees protected from predators by its uncanny resemblance to the leaves of a climbing plant.

broken and the individual bugs fly off in different directions. Flower mantises probably represent the most macabre adaptation of any insect. Disguised as a beautiful pink or white flower these insects sit and wait for some unsuspecting bee or fly to visit them, then in a fraction of a second the powerful forelegs snare their prey. This beautiful disguise also must afford the mantis itself protection from insectivorous birds.

The most familiar means of camouflage is shown by a large number of moths that when resting are indistinguishable from the bark of certain trees. Each species tends to be limited in its shape, colour and patterning to a particular species of tree. For example the wings of the pine hawk-moth (*Hyloicus pinastri*) are grey with rounded markings imitative of the scaly bark of conifers while those of the adult willow beauty (*Boarmia gemmaria*) bear a delicately traced pattern resembling the fissured bark of willow trees. Those curious looper caterpillars (larvae of geometrid moths) that often turn up during the summer crawling up one's hand, are when seen on a branch a perfect replica of twigs – securing themselves by their clasper and standing rigidly out at an angle. When the adult buff-tip moth (*Phalera bucephala*) rests on the ground beneath a tree it looks exactly like a small broken twig – the scars are simulated by light patches of scales on the head and end of the forewing.

The peppered moth (*Biston betularia*) is now a classic example of industrial melanism – the insect has evolved dark forms in industrial areas to match the soot-coated trunks of trees. In unpolluted areas melanistic forms are rare; the normally pale moth with fine black markings blends in perfectly with the bark of birch trees on which it rests during the day. The peppered moth is considered just one of hundreds of insects that have adapted in a very short time to man's activities.

It is the great potential that insects have for change which is incredible – evidently there are insects that in Africa have adapted into black forms in areas where forest fires are prevalent. Grasshoppers have been known to change from bright green early in the season to match the young grass to pale yellow later on when the grass has become dry and parched. The range of camouflage and disguise exhibited by insects is phenomenal and all associated with one common theme – survival.

Left: The leaf-like form of the brimstone butterfly (*Gonepteryx rhamni*) affords it protection when resting on cool wet days.

Below: Alluring coloration shown by the Malaysian flower mantis (*Hymenopus bicornis*). The petal-like expansions of the body and legs attract flower-visiting insects which are quickly seized and eaten.

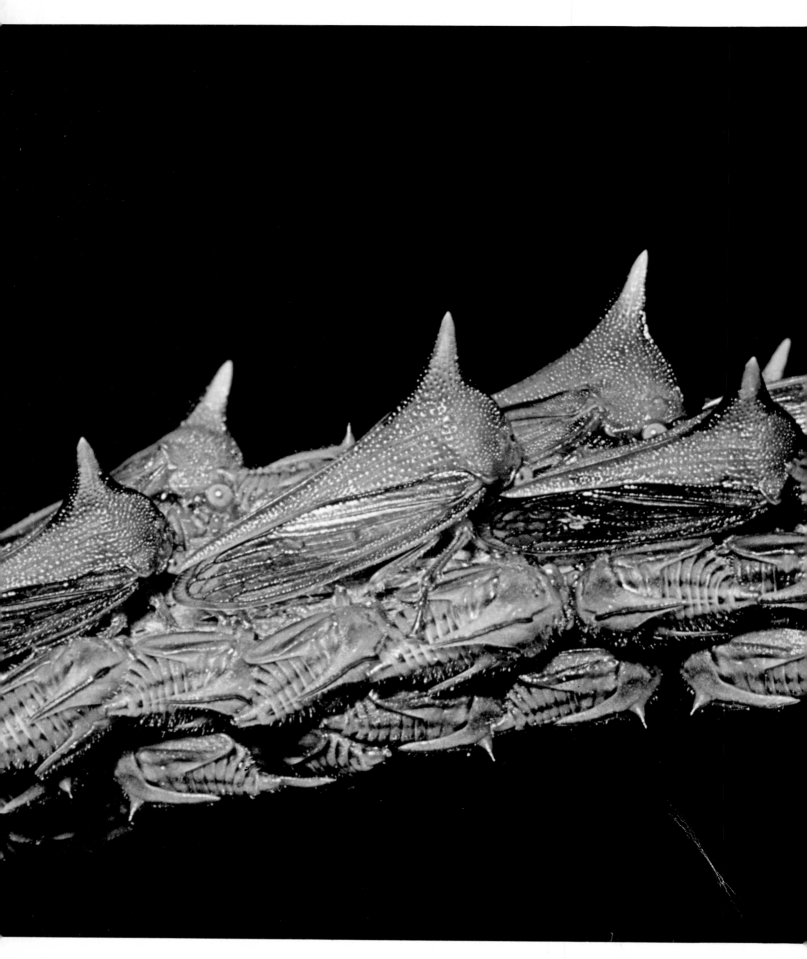

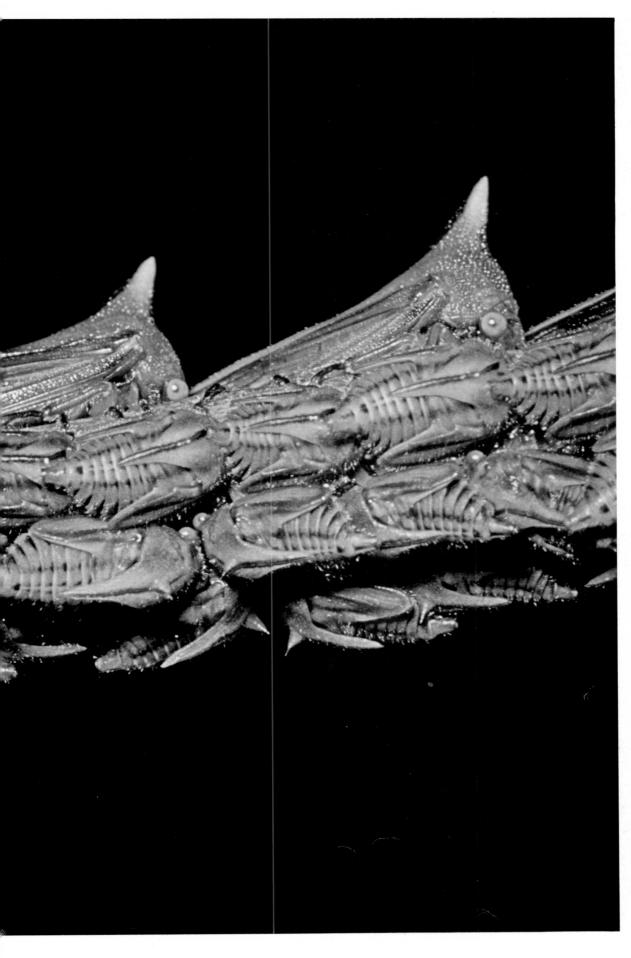

Left: Thorn-tree
hoppers lie packed
together along a vine.
The adult insects
resemble thorns
while the brown
nymphs resemble
rough bark.

Snake Mimics

Coloured for survival

The most vividly coloured reptiles are the coral snakes of the New World, comprising the three distinct genera *Micrurus*, *Micruroides* and *Leptomicrurus*. All three genera have striking red, yellow or white and black rings along their bodies similar but even more distinct than the kraits of Asia. These bright colours are common warning colours found, apart from reptiles, in the insects especially among bugs, beetles and butterflies to advertise their distastefulness and in the case of coral snakes their highly toxic venom.

Throughout the coral snakes' range are other snakes that have the similar bold patterning of the coral snakes, including the harmless milk snake (*Lampropeltis doliata*). In the western states of North America young individuals up to 1 metre (40 inches) show the bold coloration but above this size are completely black. In more northern parts of their range where no coral snakes occur local sub-species of the milk snake do not display gaudy, conspicuous skin colours and patterns. One would think that this similarity in body coloration was a simple case of mimicry, the harmless snake being afforded protection against predators by having similar patterning to the more venomous species. This may be the case, but usually in what is called Batesian mimicry, where a harmless creature mimics a dangerous one, the model is more numerous that the mimic and in this case wherever they occur together in the same habitat the population of milk snakes exceeds that of coral snakes. Also, the poison coral snakes possess is extremely virulent, a powerful neurotoxin similar to that of cobras. An animal bitten by one of these snakes would be almost certainly killed and have little chance of passing on its experience to future generations. The herpetologist Robert

Below: In the western United States the harmless long-nosed snake (*Rheinochelus lecontei*) illustrated, mimics the venomous Arizona coral snake (*Micruroides euryxanthus*).

Mertens postulated another theory which regarded both the milk snake and the highly venomous coral snakes as mimics of a third group of moderately poisonous coral snakes which include the highly aggressive species *Erythrolamprus aesculapii*. The painful but not fatal bites administered by this species would teach the victim not to associate with all snakes having a similar body pattern.

One requirement of this theory is that animals when rooting around for food are bitten by the moderately poisonous snake, a situation which has so far been difficult to prove. It may just be that all animals have evolved an innate fear of these snakes, their knowledge being rooted in instinct rather than learned.

Species and Behaviour

In the majority of coral snakes the only teeth in the upper jaw are two small poison fangs, and though strong in action the total yield of poison is low. For example the Eastern coral snake of the south-eastern U.S.A. (*Micrurus fulvius*) contains only 2.6 milligrams (0.00009 ounces), the maximum yield being up to 200 milligrams (0.007 ounces) for the Brazilian *M. corallinus* which together with the related *M. frontalis*, known locally as the 'cobra coral', constitute the most dangerous species. Fortunately these snakes do not have a wide gape and normally attack only small prey including other snakes, lizards, frogs and young birds. Once the prey is secured the snake 'chews' its victim several times, each time injecting a dose of venom.

One peculiar habit of coral snakes, also shown by some species of milk snake, is their defence posture – when threatened the small head is hidden within the body coils and the short blunt tail is raised several centimetres and waved to and fro giving the appearance of a substitute head. Presumably this is a device, like the lizard's trick of casting its tail, of distracting a predator such as a hawk or eagle, making it lunge at a less essential part of the body, giving the snake a split-second chance to make its getaway.

In North America there is a well-known colour rule to distinguish which snake is or is not dangerous. The dangerous species always have the yellow or white ring bordered by red, whereas in the harmless colubrid the red and yellow are separated by a black ring. This rule does not apply south of the border

Above: the non-venomous milk snake (*Lampropeltis doliata*).

Left: Almost indistinguishable – a deadly Brazilian coral snake (family Elapidae), top, compared to its non-poisonous mimic, a member of the family Colubridae, below.

where both groups of snakes are more numerous. Mexico has at least thirty different species of coral snakes and Brazil contains the largest species *M. spixi* which grows to 1.5 metres (5 feet) long. Most species are much smaller, on average 60 centimetres (2 feet) or less.

The genus *Leptomicrurus* consists of a group of extremely slender coral snakes which occur from Colombia as far south as Peru.

The coral snakes are in reality an enigma for although they advertise themselves with their shiny bright markings, most of them live a large part of their lives underground tunnelling amongst leaf litter or laying up in the abandoned burrows of rodents.

The Octopus
Chameleon of the deep

One of the most fascinating features of the octopus is its ability to undergo rapid and startling colour changes. Even within its egg capsule the young octopus can show remarkable flushes of colour even though it has a complement of only around seventy colour cells. By the time an octopus reaches adulthood the number of colour cells has increased to between two and three million which gives the octopus a delicacy of colour unrivalled in the animal kingdom.

Its basic equipment consists of groups of cells called chromatophores which can be likened to tiny elastic bags filled with coloured pigments. When the cells are expanded by contraction of radial muscles attached to the cell like the spokes of a bicycle wheel, the pigment is distributed over a wide area and the colour it contains dominates the octopus's body colour.

Colour cells

When the chromatophores are contracted by the natural elasticity of the cell the pigment is concentrated into a tiny speck. Octopus are thought to have two main sets of colour cells, a larger darker set containing black-red pigment and a smaller lighter set containing orange-yellow pigment. Beneath the chromatophores lie other cells that add to the range and quality of the colour produced. One group of cells called iridiophores contain reflective material and are used by the octopus to camouflage itself by matching the quality of reflected light of its surroundings. A further set of cells known as leucophores reflect white light and when they are dominant the whole of the octopus's body has a pale, bleached-out appearance.

The kaleidoscope of colour shown by one specimen visited periodically at one spot on the coral reef is described by the naturalist Gilbert Klingel in his book *Inagua* which describes the wealth of marine life of the Caribbean Sea:

> It always seemed irritated by my presence. Its nervousness may have been caused by fear, for it certainly made no pretence of belligerency, and it constantly underwent a series of pigment alterations that were little short of marvellous. . . The most common colours were creamy white, mottled vandyke brown, maroon, bluish grey and finally light ultramarine. . . When most agitated it turned lived white, which is I believe, the reaction of fear. During some of its changes it became streaked at times in wide bands of maroon and cream, and once or twice in wavy lines of deep lavender and rose. Even red spots and irregular purplish polka dots were included in the

Below: Octopuses can change their skin-texture as well as colour to match their surroundings.

repertoire though these gaudy variations seldom lasted long.

Not only can an octopus change colour, it can change the texture of its skin from a velvet smoothness to a corrugated warty appearance which blends in with weed-encrusted rocks and coralline growths. Young octopuses often become crenellated in form when threatened. This description by another naturalist, Joseph Sinel, describes both the colour and textural changes of an adult.

When highly content, as after a meal, and perched, as it is fond of perching on an eminence, the papillae (pimple-like projections of the skin) are erected, and these are always of an orange colour. Oftentimes the whole body will be marked off in irregular honeycomb patches, or more like crocodile skin. First some of the patches are purple, others orange, then these colours are reversed. When danger threatens, or even when the hand moves towards it . . . the animal winces, and turns an ashy-grey.

Both authors were correct in their description when they noted that a frightened octopus takes on a blanched appearance. Under experimental conditions a frightened octopus flattens out, becomes ghostly pale and often the effect is heightened by a darkening of the chromatophores around the eyes giving the whole animal a macabre aspect.

All colour and textural changes are under nervous control. An octopus is well endowed with nerve cells having some 168 million cells concentrated in the brain, which forms a ring around the oesophagus behind the horny beak and between the eyes. There is a visual centre in the brain which interprets stimuli received via the eyes. Another centre interprets tactile stimulation sent back to the brain by nerve fibres in the tentacles. The whole machinery is extremely complicated and the octopus is considered one of the most advanced and intelligent of invertebrate animals, having a wide capacity for learning.

Octopus have been much studied by marine biologists and subjected like mice and rats to the full gauntlet of physiological tests involving tasks and rewards. It is no wonder that some octopus make desperate efforts to escape and will climb out of their tanks or, probably unwittingly, remove the rubber bungs from them. As a result of behavioural studies the common octopus (*Octopus vulgaris*) especially has been shown to display specific colours and patterning in response to specific situations. When about to rush a crab the common octopus will suddenly change from its normal yellow-grey colour to bright vermilion. If the octopus wants its prey to move into a better position for attack its body becomes suffused with moving waves of colour which are thought to stimulate the prey into movement.

Patterns of Conflict

When confused and in a conflict situation between attack and retreat, the body becomes mottled. During courtship a specific striped patterning of the limbs is shown by the related *O. cyanea*. These contrast markings are even more noticeable in the related common cuttlefish (*Sepia officinalis*) which even surpasses the octopus in its speed of tonal change.

Above: Octopuses employ rapid colour change to denote emotions – in this case as a dramatic form of warning display.

143

Coral Reef Fishes

Marvels of marine pageantry

The warm, shallow waters of coral reefs abound with life that is more diverse and colourful than that of any other part of the ocean. The living corals themselves may give patches of greens, beiges and browns as a background to the greater brilliance of many of the inhabitants. Dead coral has little colour, but provides surfaces for the growth of algae, shelter for fishes and their invertebrate neighbours. The whole reef forms one of the most gorgeous and fascinating ecosystems in the world.

Many of its inhabitants, such as the scorpion fish (*Dendroscorpaena cirrhosa*), have evolved camouflage that makes them virtually invisible while they lie motionless on the seabed waiting for a careless fish to swim near enough to be caught after a swift charge. The main characteristic of life in the waters of the reef is quite opposite to the scorpion fish's low profile: it is an apparently rash display of ostentatious colour and form.

Fishes that do not rely on remaining hidden have various defences: they may be able to eject a powerful and repulsive smell, sting an attacker, deliver a poisonous bite, or the fish may taste nasty and its flesh may be poisonous. The deterrent value of these defences is fully effective only when a potential attacker knows what to expect if he presses an attack. It would be pointless if a fish were to poison its assailant only after it had been eaten, so these defences are well advertised by the beautiful markings of many of the reef fish. These defences are not foolproof; several predators can overcome them or have developed an immunity to them – the hammerhead shark seems to be undismayed by the spines of the stingrays which it eats whenever it can – but the defences are successful enough to have encouraged their evolution in many species of fishes found swimming among the corals.

It is perhaps curious that the marvellous flashing colours darting among the corals and seaweed of a reef are probably poorly perceived by the fish they are most intended to deceive. Many large fish that normally live in fairly deep water have poor colour sight. They have evolved eyes that see well in the nearly monochromatic conditions of their usual habitat: most red light is filtered out of the spectrum at a depth of 10–20 metres (33–66 feet), red fish appearing nearly black, and orange and yellow disappear a little lower. Large predators may have only poor colour appreciation, but they can be deceived by the tonal pattern the colours transmit. The lion fish (*Pterois volitans*), also called the red–and–white sea dragon, looks a magnificent sight to us, and it is likely that fish which live near to the surface – and whose colour perception is much better than, for example, a shark's – see the warning blaze of colour. A large predator may see the red stripes merely as a dark pattern which makes it difficult to recognize the fish's shape: even its eyes are camouflaged by the 'dark' stripe. If the warning and the camouflage effect fail to discourage the predator, the lion fish's venomous sting, sited in its dorsal fin, may save it from serious injury.

The clown fish (*Amphiprion sp.*) declares its presence with a striking pattern of orange, black and white. It does not appear to have any defence of its own, but lives in a well defended territory. Its body is covered with mucus that protects it from its 'host', the large sea anemone, *Stoichactis,* whose tentacles sting and paralyse other fish. The clown fish's colour may act as a lure to the sea anemone's prey, but nothing is certain on the subject of what the sea anemone gets out of this colourful association, if anything at all.

Sea anemones unintentionally supply the venom that forms the armoury of one of the most beautifully coloured of all sea creatures. In some undiscovered way, the sea-slugs (such as (*Flabelligera iodinea*) consume sea anemones without triggering their stinging cells, which the sea-slug digests. They make their way from the sea-slug's digestive tract, through the outer tissues of its body until they appear as fronds of stinging cells along the sea-slug's back, where they discourage any predator who chooses to disregard the colours that advertise its poisonous nature.

One of the remarkable uses to which some

Left: A blue surgeon fish (*Paracanthus theutis*), one of over seventy-five species from tropical waters. The fish are so named because of a pair of sharp scalpel-like spines located on each side of the tail.

Right: A pair of rabbit fish (*Siganus*) in the Red Sea. Since the Suez Canal was opened, rabbit fish have made their way into the Mediterranean as far as Greece.

Right: A young four-eye butterfly fish (*Chaetodon capistratus*) from the Caribbean.

reef fishes put their colour is found in the way *Diploprion drachi* approaches its prey. This small, pale blue and slightly hump-backed fish chooses a large fish of similar colour and uses it as a 'stalking horse'. Swimming close to it, almost pressed against its body, the small predator sometimes looks as though it is part of the larger fish. When it is close to its prey, *Diploprion* makes its attack on the small fish that had noticed only the large fish that had been swimming by with no implicit threat. *Cheilinus diagrammus*, a small wrasse, stalks its prey in a similar way but showing yet more ingenuity. It changes colour to match the nearest 'stalking horse', whether it is the purple of the purple parrot fish (*Scarus niger*), a soft grey to blend with the sandy coloured parrot fish (*Hipposcarus harid*) or a mottled orange to help it merge with the corals of the seafloor.

This form of mimicry is carried a step further when some predatory fish assume the colours of harmless herbivores, such as the common surgeon fish (*Acanthurus sohal*) and the exquisitely coloured yellow butterfly fish (*Chaetodon semilarvatus*). They swim with the group of herbivores until they are in range of a suitable prey. *C. diagrammus* is especially adaptable in this way. It can even reproduce the lateral stripe as well as the general colour of the goatfish (*Parupeneus macronema*). Another wrasse, the sling jawed wrasse (*Epibulus insidiator*), so-called because of its capacity to protrude its mouth into a long tube with which it can hoover small creatures from the shelter of tiny crannies in the coral, is usually a blue fish with an orange-yellow band across its shoulders. When hunting, it may turn completely black. Females of the species associate with the yellow butterfly fish.

These wonderful colour changes occur as pigmented capsules in the fish's skin expand or contract. The fish can control their thickness, and therefore their colour, at will, by means of minute muscle cells that are closely associated with the capsules.

Many of the small fishes of the reef move in shoals for safety from predators. When one attacks the shoal, the swirl of brilliantly coloured fish makes a distracting and scintillating cloud that confuses the predator. It is important that these small fish can recognize their species

rapidly so that the shoal can function. In this situation, their colours are worn rather like those of a football team, whose players can recognize the barest glimpse of a teammate's shirt and socks. Among these shoaling fish are the damsel fish (family Pomacentridae) which browse the plankton: one of the smallest fish to feed on this material.

The cleaner wrasses, of which *Labroides dimidiatus* is a well-known example, are clearly striped in blue and white. They approach even large predatory fish to nibble away any extoparasites and growths on their skins. The wrasse does not depend entirely on its shape and colour to give it immunity from the predator's appetite, it swims with a characteristically undulating action. Unfortunately for the fish that is being cleaned, the sabre-toothed blenny (*Aspidonotus taeniatus*) looks rather like the wrasse. It, too, imitates the wrasse's swimming motion, and approaches to nibble at the skin of the larger fish, but tears away a piece of fin or another enjoyable morsel.

Night and Day

The reef's occupants work on a kind of shift system. They change places at dawn and dusk. The night fishes are a different group to the daytime fishes. As the damsel fishes leave the upper levels of the reef's waters after a day's feeding, they swim down to the shelter of the corals below, and their place in the reef's activity is taken by the brilliant red cardinal fish (*Apogon sp.*). Their glorious colour turns to nearly black in the darkness, making them nearly invisible. Other fish change colour with the fading of the light. The French grunt, which is a shadow-striped sulphur in daylight fades to a pale grey, patched with greyish brown that blends beautifully with the corals of the seabed at night. The darkness is lit with luminescence here and there, the most brilliant of which may come from the *Photoblepharon* fish which has a pea-sized organ below each eye. In this bacteria produce a chemical light, although for what reason no one yet knows.

In the waters of the coral reefs, the system is as full of serenity, violence and beauty as any part of the world of nature. The strong role played in it by the inhabitants' use of colour is something we must see with thrilled gratitude.

Nightjars, Frogmouths and Potoos
The feathered deceivers

For birds, vigilance is a vital part of the state of being alive. They learn its importance early and survive by its innate discipline. There are few moments when a bird, even when resting, preening or feeding, does not look up and about every few seconds, searching its surroundings for a threat. Many of the more vulnerable birds – those which nest in accessible places or ones that cannot fly well – have evolved effective passive defences by using the cryptic pattern and colours of their plumage as a camouflage. It would scarcely be possible to find more accomplished expression of this form of defence than in the birds of the order Caprimulgiformes.

Nightjars
In Europe, the nightjar (*Caprimulgus europaeus*) is rarely seen when it is not in flight. It haunts the fringes of woodlands and heath, flying on wings that are as silent as an owl's in search of insects early in the morning and at dusk. The nightjar's plumage is streaked, barred and mottled with reddish and dark brown and with pale buff. The background colour is a soft grey-brown. When the bird rests on the ground, its tail lying straight back and low and its wings folded so that their tips project backwards and a little upwards, it merges perfectly with the broken textures of the surrounding woodland floor or the heath. Frequently, the birds choose a rest or a nesting site where there is a scatter of rotting wood, against which their plumage is nearly invisible. It is even harder to perceive the bird when it flattens itself against the trunk of a tree when alarmed. It generally aligns itself with the pattern of the bark on a trunk or branch, only rarely perching across the branch rather than along it.

Like all birds that must rely on their flight for food, preening is important to the nightjar. It takes some of its food from the ground, but much of its insect prey is caught on the wing. Its wide mouth and the 'powder puff' of fine bristles around its gape scoop up insects in

Right: The cryptic coloration and body form of the tawny frogmouth is one of the most effective examples of camouflage among the larger animals. At night the frogmouth becomes active taking most of its insect food in the trees and on the ground.

ghost-quiet swoops after pauses when the bird hovers in search of its prey. If the preening were done in daylight, the effect of the camouflage would be undone, but the nightjar preens at night or at dusk. It may preen the delicate bristles round its bill while in flight, using a group of serrations on the middle claw of its three front toes.

The nightjar does not build a nest. It scrapes a hollow in a suitable place and incubates its eggs there. The site is chosen with care, giving the sitting bird a good all-round view of any danger that might approach. She relies on her superb camouflage for the safety of herself and her eggs. If she or her mate find their deception penetrated, they will fain injury, either on the wing or on the ground, in an attempt to draw a predator away from the scrape with its precious contents of eggs or chicks: but the motionless defence of her camouflage will be maintained until the last moment.

When the season changes to winter, and frost and snow whiten the ground, the nightjar flies south to find a more suitable hunting ground and, incidentally, a more fitting background to its plumage. It makes a winter home in Africa until spring warms Europe to life again.

Frogmouths

While the nightjar is master of camouflage in the heaths and broken woodlands, the frogmouths have mastered the art of invisibility in the trees of Australia, Papua-New Guinea and Asia. They are so successful that ornithologists are fairly sure that several species may still be evading their sharp eyes in the jungles of southern Asia. These Asian birds are of the genus *Batrachostomus,* and include the large frogmouth (*B. auritus*) and eight other but much smaller and far less well-known frogmouths. The Australian genus, *Podargus,* consists of three species, one of which (*P. strigoides*) is found only in Australia.

The frogmouths carry the art of camouflage one stage further than the nightjars. They do not rely wholly on the colour and pattern of their plumage, but adopt a strange and effective pose which makes them yet harder to detect. They stretch their bills so that their necks and bodies are in a straight line, press their tails close to a branch and seem to stiffen into the appearance of the wooden material they perch on. The upright position of the bill, surrounded

by the fine rictal bristles, gives the impression of an abruptly broken branch. Like the nightjar, a frogmouth will 'freeze' when approached by a predator, but at the last moment, it may open wide its brightly coloured mouth in an enormous gape, which gives some predators a nasty shock. If they believe they have been moving towards a piece of dead wood, the sudden opening of a large orange-yellow gape must be pretty unsettling.

Frogmouths build their nests in trees. The nest is an untidy structure of twigs against which the bird's plumage is almost indistinguishable. The style of its tactics in defence are matched by those of the hunt: camouflage works two ways. While it can save a bird from detection by its enemies, it can also make it invisible to its prey. The frogmouth was for many years believed to hunt on the wing like the nightjar, but its short tail reduces its agility in flight, making it a poor imitation of its European cousin. The frogmouth has powerful feet that are well suited to climbing and walking among trees, and has developed a different system of hunting. It perches, still as

Above: Mother and young tawny frogmouths (*Podargus strigoides*), a species widespread through the Australian continent.

a dead branch, even its eyes are closed to mere slits, and watches until a small reptile or an insect wanders into range. Then the bird swoops to the ground or walks along the branch to seize it.

In a country like Australia, where trees are quite often burned, the frogmouth melts out of view against the charred and broken branches, and can use this ambush technique of hunting most effectively. Its stillness when watching has often led observers to think that the bird is torpid; but if a person walks slowly round the bird's perch at a distance that does not alarm it too much, he will notice that the frogmouth's eyes follow him round. If approached too close, the bird will fly away quickly enough.

The nests of the *Batrachostomus* genus are much more ingenious than those of their Australian cousins. The Ceylon frogmouth (*B. moniliger*) mixes bark and fragments of moss with down plucked from its breast to form a lump that it presses onto the branch of a tree. This precarious structure looks like a knobbly part of the tree, and from its scant protection the bird hatches and rears her young. The pale-coloured hatchlings are unprotected by the coat of invisibility worn by their mother so the adults cover and guard them until they are fully fledged.

Potoos

In Central and South America, a motorist can be startled to see the top fly off one of the endless rows of fencing posts that line the sides of the roads. The common potoo (*Nyctibius griseus*) will join itself to a post rather than perch on it. Its tail pressed down its length, and its bill – opened to give the illusion of squareness – pointing to the sky: it looks like an extension of the post. Birdwatchers in these regions look for a post that is slightly longer than its neighbours, and find their potoos in this way.

The bird has finely marked plumage that exactly matches the insect-riddled wood of posts and rotting branches. Like the frogmouth, the potoo hunts from its perch, but it flies well enough to catch larger insects on the wing. In fact, its lifestyle falls neatly between that of the frogmouths and the nightjars.
The Caprimulgiformes have mastered the secret of leading hidden and effective lives in a large part of the world. All they need is debris and trees for safety, and insects for food.

Above: The common potoo (*Nyctibius griseus*) of tropical South America hunts at night taking large insects by flying after them from a low branch or post.

Above: A European nightjar (*Caprimulgus europaeus*) is perfectly camouflaged as she incubates her two eggs. The nightjar, a summer visitor to Europe, nests in open woodland and heaths and preys on a variety of flying insects at dawn and dusk.

Birds of Paradise
Plumes of splendour

Far right: The fabulous plumage of the birds of paradise is shown to its best advantage during courtship when the gaudy male birds (illustrated) attract the females. When posturing, patches of feathers normally hidden are suddenly revealed – bright metallic throat pouches or tail plumes which form glorious cascades of colour.

Survivors of Magellan's circumnavigation of the world first opened European eyes to the heraldic plumage of birds of paradise. Later seafarers brought more skins of the gorgeous birds to the northern hemisphere. The method used in skinning them made it appear that they had no legs, a deception that gave birth to many tall stories about them. It was said that they flew constantly, not landing even to nest. The myth-makers told how the female laid her eggs in a hollow in the male bird's back, where she incubated it while the male flew about the forest with his extraordinary burden. From the beginning, scientific naturalists scorned these tales, but it was not until 1824 that an educated European observed the birds in their native Papua-New Guinea. He was able to report that these were celestially beautiful birds indeed, but capable of grasping branches with strong feet.

Birds of paradise are close relatives of the crows, with their powerful feet, stout bills and strong bodies, and they vary in length of body from 17–120 centimetres (7–47 inches). About twenty genera and forty-three species of these splendid birds live in Papua-New Guinea, the Moluccas, north-eastern Australia and some nearby islands. The isolation of some species leaves us ignorant of their habits, and there may well be some undescribed members of the family Paradisaeidae living in inaccessible areas of the mountains and rainforests. The birds live on fruit and berries principally, but some species will also eat insects, frogs, tree lizards and other small vertebrates. Their preferences vary from species to species, but none is able to live on their legendarily exclusive diet of dew. Their habitat ranges from the dense rainforests of the lower lands of the region to high mountain forests above 3000 metres (10,000 feet). A few, such as some red-plumed birds of paradise (*Paradisaea raggiana*), live close to cultivated fields, and display their plumes in the trees that grow on the outskirts of villages.

Plumage and Display
It is the extravagantly varied and splendid plumage of birds of paradise that won them their grand name. Many of the birds have brilliantly coloured feathers on their bodies and wings, but other species are mostly black with small, jewel-bright areas of colour. The most remarkable effects are produced by the pennant-like plumes that 'fly' at the ends of nearly bare vanes. Other feathers are minute and so brilliantly iridescent that they seem to be enamelled on polished metal. In contrast, cascades of long, gaily coloured feathers spread over some birds' bodies. Perhaps the rare Waigeu bird of paradise (*Diphyllodes respublica*) has one of the strangest appearances. Its back and wings are red, with a mantle of yellow, and its breast is dark green, but the crown of its head is naked. This bare skin is of a bright cobalt blue and is patterned with delicate lines of small black feathers.

It seems that the impressive colours and flamboyant styles of the plumage of birds of paradise are not enough to satisfy their need for outward show. They display them in many ways to attract their mates. The male Waigeu bird of paradise ensures the full effect of his striking colours by plucking all the leaves and shoots from the branches that surround his chosen perch, and by clearing a circular area of the ground below. Only when he has prepared his 'stage' does he begin his display to attract a mate.

Birds of paradise rarely congregate. They live solitary lives except in the breeding season, when they fly to their arenas or stages. These are areas the birds use for their courting displays, and the cocks secure a patch of this territory for themselves and defend it fiercely. The strongest and most aggressive occupy the territories in the middle of the stage. Once these territories have been established, the displays may commence. With some birds this is an acrobatic event. The blue bird of paradise (*P. rudolphi*), a middle-sized bird of the mountain forests, begins his show by calling for attention. When he sees possible mates are near, he grasps his perch and in slow motion rolls backwards until he is hanging, bat-like, upside down. With a flutter, he opens his wings and shakes out his plumage to form a fan of

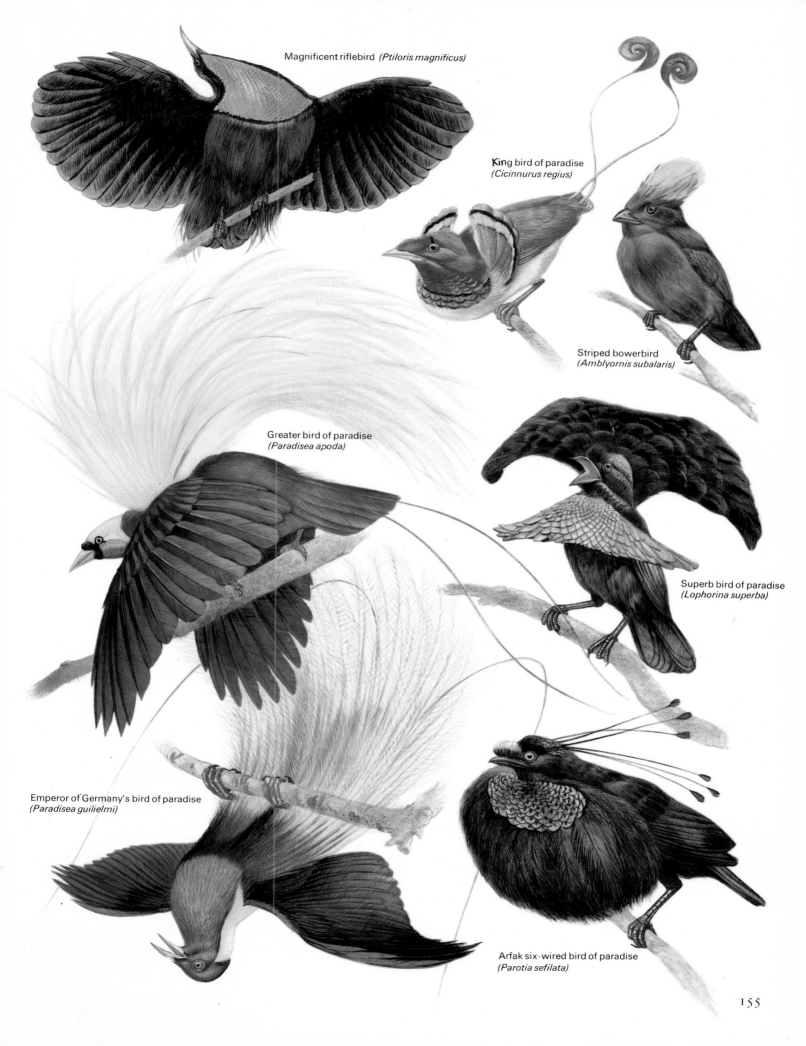

Magnificent riflebird *(Ptiloris magnificus)*

King bird of paradise
(Cicinnurus regius)

Striped bowerbird
(Amblyornis subalaris)

Greater bird of paradise
(Paradisea apoda)

Superb bird of paradise
(Lophorina superba)

Emperor of Germany's bird of paradise
(Paradisea guilielmi)

Arfak six-wired bird of paradise
(Parotia sefilata)

Right: The 15-centimetre (6-inch) long king bird of paradise (*Cincinnurus regius*) is the smallest of the forty-three known species. It is mainly found in the lowlands of western Papua-New Guinea.

iridescent blue with his breast, in contrasting bands of colour, as its centrepiece. Above the open fan spring a pair of wire-like feathers that gracefully arch to right and left, and end in 'eyes' of brilliant hues. Firmly gripping his branch, the bird begins a jigging dance from its hip joints. The effect is to alter the iridescence of his feathers and to create a display of changing colour that must be alluring to any female of his species. He accompanies the show with a soft repetitive song.

A privileged few observers have seen the stages of these displays in full use. As many as twenty birds may use a group of trees as their stage, and display their glorious plumage at one time. Alfred Russell Wallace in 1869 published the earliest description of the display behaviour of the Aru greater bird of paradise (*P. apoda apoda*), telling how the birds constantly vibrate their long, golden fans into arching plumes as

they fly from branch to branch of a display tree, pausing to perch, raise their wings upright above their backs and bend their heads down and forwards. This whole chorus of gold, red and delicate buffs moved the naturalist greatly. Some species, such as the red-plumed bird of paradise, are so enraptured by their dance that they become rigid and immobile at the peak of their demonstration.

In the dense treetops of the lowlands of Papua-New Guinea, lives the king bird of paradise (*Cincinnurus regius*), smallest of the family. About the size of a starling, its plumage is of an intense cinnabar-red, which has been said to shine like spun glass. This colour alters to an equally intense orange on its smooth head. A strip of iridescent green separates the red of its throat from the white of its breast, and a small patch of the same green is clearly defined above each eye. Even its legs are coloured, an exquisite cobalt. Yet the mate of this jewel of the forest is an unremarkable brown bird, whose colour fades a little on its underparts. She bears none of her mate's pennant feathers. These spring from the two middle feathers of the king bird's tail as bare, spiralling vanes that terminate in a pair of brilliant 'eyes' which hang about 12.5 centimetres (5 inches) below its tail.

In most species, the female is a drab consort for her gorgeous mate. Observers have noticed that where this distinction is strongest, the cocks lose interest in the females directly the breeding season ends. In species where the distinction is less apparent, the cocks may help to construct nests and, more rarely still, feed the young. Most birds of paradise build large, bowl-shaped nests in trees, but there are exceptions. The king bird of paradise nests in a hole in a tree, and the multi-crested bird of paradise (*Cnemophilus macgregorii*) builds a nest with a roof, selecting a spot near the ground. It is, perhaps, surprising that despite the brilliance of the male birds' displays some hybrids occur in the wild. These are generally sterile. The commonest are probably crosses between the king bird and the magnificent bird of paradise (*Diphyllodes magnificus*). It is unfortunate that there is too little understanding of birds of this family to enable ornithologists to evaluate the likelihood of genetic mutations taking place which might produce new and stable characters in the family. Hybridization occurs most commonly among birds that have the most gorgeous plumage, suggesting to some orni-

Left: Count Raggi's bird of paradise (*Paradisaea raggiana*) is one of several species whose generous tail plumes are used by the tribesmen of Papua-New Guinea to decorate their elaborate head-dresses.

thologists a striving towards a means of avoiding these fruitless unions.

Near relatives of the birds of paradise, and near neighbours too, are the bower birds (of the sub-family Ptilonorhynchinae). These intelligent and fascinating birds depend on intriguing their mates by building some of the most interesting and beautiful structures made in the animal world. They have less impressive plumage than the birds of paradise, but the colour and shapes of their courting areas and nests makes up for any lack of colour in their feathers.

Exploitation

The brilliance and beauty of the true birds of paradise have brought them much danger from man. For thousands of years the human population of the birds' homeland has hunted them for their feathers. The men of Papua-New Guinea collect the plumes and construct the most magnificent regalia for feast days and war parties. At one gathering, they may be wearing feathers worth nearly £200,000. The bird population in the past did not suffer unduly in providing the materials for this human display. Families of men inherited the rights to certain display stages, and they would kill only mature birds after the mating had taken place. The trade with Western nations changed this, and the shotgun wrought great havoc among the birds until unlicensed killing or possession of birds of paradise in Papua-New Guinea became illegal in 1924. The birds are not yet safe from this danger because more of the local population – who are not banned from hunting them – have shotguns and, with fewer wars being fought, can wander further afield to discover new breeding stages. There are also commercial interests which pose a threat.

157

THE USES OF COLOUR

Name	Colour patterns	Behaviour associated with colour use
Blue Bird of Paradise *Paradisea rudolphi*	Both sexes are black with blue wings. The adult male has long blue purple and cinnamon plumes on its flanks, and two very long black tail streamers.	Male swings upside down from a branch at a selected display site, shaking out the iridescent and brilliant flank plumage. He sings a soft monotonous song and moves the hips to shake out the plumes and tail streamers.
Frigate bird *Fregata magnificens*	Blackish-brown with a metallic sheen in the male, which also has an orange colour sac beneath the throat which turns bright crimson in the breeding season. The female plumage is dull.	The brilliant red throat pouch is inflated by the males, while raising and shaking the head and calling. These birds nest in colonies on ocean islands.
Indian peafowl *Pavo cristatus*	Royal blue, greenish blue and bronze-green head and body for male, with a train of feathers supported by the tail, metallic green with 'eyes' of bronze and purple. Female mottled brown, buff.	The peacock displays the magnificant tail by spreading it out and raising it over his head announcing the display and his ownership of the territory by screaming. The tail is shaken to produce a rattling noise.
Potoo *Nyctibius grandis*	Dark brownish, buff and black speckled bark-like pattern.	The potoo lives in South American rain forests where it perches vertically on the end of dead tree stumps with tail pressed against the wood so that it appears to be part of the tree.
Coral snake *Erythrolamprus aesculopii*	Banded with alternating rings of red, yellow and black.	No special display is given by this species, though some coral snakes will wave their tail around in a head-mimicking display.
King snake *Lampropeltis polyzona*	Banded with alternating rings of red, yellow and black, almost identical to previous species.	No special display.
Chameleon *Chamaeleo vulgaris*	Shades of brown, green, buff and yellow, changing continuously to match colours of the background.	Very slow movements help the chameleon to merge with the background.
Plaice *Pleuronectes platessa*	Mottled pattern of browns, buff, black and grey. The body is flattened and the fish lies on its left side and the left eye migrates during development to join the right eye on the upper surface.	Swims close to sandy or gravelly sea floor. When at rest, shakes sand over its body to blur the outlines and conceal itself even more effectively.
Hornet *Vespa crabro*	Abdomen marked with yellow stripes on a brownish-black background.	No special behavioural adaptations.
Mantis *Choeradodis rhomboidea*	This preying mantis is bright green, with a flattened thorax and abdomen, patterned with leaf-like veining, and closely resembles the vine leaves it lives among.	Very slow moving, the mantis stalks its prey so as to keep as well-concealed as possible.
Common octopus *Octopus vulgaris*	Basically pinkish-brown in colour, this mollusc is continuously changing shade, with a repertoire from dark brown to white via reddish and orange. Can alter its hue almost instantaneously.	Violent 'blushing' displays when provoked. If attacked some octopuses will discharge an ink cloud to distract the predator, turn almost white, and jet quickly away.
Sea Slug *Facelina auriculata*	One of the larger British sea slugs, this animal has brilliant crimson and white markings on the finger-shaped processes called cerata which cover its back. It has no shell.	Sea slugs of this group feed on sea anemones and other stinging coelenterates, but the stinging cells instead of being discharged are taken into the gut and then stored in the slug's cerata, and will be discharged and sting any fish that bites it.

Main effects of colour patterns

Sexual display (male)

Sexual display (male)

Sexual display (male) Camouflage (female)
The peahen alone broods the eggs and cares for
the young, sheltering in grass or scrub.

Camouflage
Potoos and nightjars are most active by night,
and their superb camouflage allows them to rest
by day undisturbed by predators.

Warning
The garish colours of coral snakes are shown by a
large number of species, and predators have learned
to avoid all of them. This species is highly venomous,
though not aggressive.

Mimicry
The harmless King Snake has the warning colours
of dangerous coral snakes to protect it from
predators.

Camouflage
Used by chameleons both for protection against
predators and to disguise them from the insects
they prey on.

Camouflage
Another set of adaptations for a different habitat and
way of life serve the same purpose for plaice as for
pipefish.

Warning
Black and orange or yellow are common warning
colours for insects; well-deserved for the hornet
with its formidable sting.

Camouflage
The resemblance to the foliage it lives among helps
this ferocious predator to catch the insects it feeds on,
while also protecting it against the animals which
might feed on it.

Camouflage
Octopuses are carnivorous, and their mastery of
colour change assists hunting and protection from
other predators. Octopuses may also communicate
with each other by colour changes.

Warning
The bright colours of many sea slugs are believed to
warn predatory fish that they are distasteful.

A SHORT LIFE
—AND—
A LONG ONE

A giant Galapagos tortoise
(*Testudo elephantopus*) one of the largest and
longest-lived of reptiles.

The Giant Tortoise

Grand old man of the reptile world

As the fable of the tortoise and the hare suggests, there would appear to be some advantage in a slow, unhurried life, for tortoises are acknowledged to be the longest-lived of all animals. They are also one of the most ancient groups of land vertebrates, having existed relatively unchanged for 150 million years.

Because of the curiosity they aroused coupled with an extremely placid nature, these animals were regularly collected by explorers who donated them, on their return, to private gardens and zoos where many have survived to a great age. The record for longevity was held by a specimen of the now extinct Seychelles tortoise brought from its home islands to Mauritius by the French explorer Marion de Fresne in 1766. Known as Marion's tortoise it lived in the artillery barracks at Port Louis until 1918 when it was accidentally killed by falling through a gun emplacement. At that time it had been in captivity for 152 years, and since it was adult when captured, it must have been between 170 and 180 years old.

Longevity is, however, not restricted to giant tortoises. A 'common' Mediterranean tortoise, placed in the gardens of Lambeth Palace in 1633 and allowed to roam freely, was said to have lived until 1730 according to one account, and to 1753 according to another. A handsomely marked Malagasy starred tortoise, allegedly presented to the Queen of Tonga by Captain James Cook in 1777, was said to have survived until May 1966. Unfortunately there is no record among the captain's copious notes of the tortoise gift, nor any record of him stopping at Malagasy en route for the Tonga Isles. When some English ladies visited Tonga in the 1930s they asked to see the venerable reptile. The story goes that the Queen ordered her butler to search the garden for it and it was carried in to the audience hall 'held like a tray in the time-honoured manner of butlers'. Calculating from dates carved on their shells, there are many tortoises which have lived for over a century, including 120-year-old American box turtles, and even European pond tortoises have lived to well over seventy years.

The family Testudinae to which all land tortoises belong, have large club-shaped feet adapted for walking over often hard and dry terrain. Apart from the thick sculptured shells, the limbs, head and tail are usually covered with hard bony scales which are a perfect defence against spiny bushes and cacti on which many species feed. When attacked by a predator, a tortoise will retreat into its shell with a hissing sound. The shell is often domed which gives added protection in that it is difficult for a predator like a large cat to get a grip on it – so the tortoise waits until danger has passed when it gradually pushes out its head and limbs and continues its slow ambling existence. The only time a tortoise changes tempo is prior to mating. Gilbert White, the eighteenth-century nature diarist, observed this in the tortoises he kept in his Hampshire garden: 'The male walking on tiptoe; his fancy intent on sexual achievement which transports him beyond his usual gravity, and induces him to forget for a time his ordinary solemn deportment.' This rather eloquent account fails to mention the male ramming the female's shell (a common practice among tortoises), which serves as a brusque signal for mating. During copulation the male climbs onto the female's back and grips her shell with his forelimbs. In the spring and summer months tortoises often engage in prolonged sexual activity which is why they are regarded in the Far East as an animal of high sexual prowess.

Darwin and the Tortoise

Most spectacular are the giant tortoises (*Testudo elephantopus*) of the Galápagos Islands which lie on the equator 1000 kilometres (620 miles) west of Ecuador in the East Pacific. Charles Darwin was the first naturalist to observe these giants which measure up to 1.5 metres (5 feet) long and weigh up to 225 kilograms (about 500 pounds). An excerpt from his diary of 17 September 1835, reads:

The day was glowing hot, and the scrambling over the rough surfaces . . . very fatiguing; but I was well repaid by the strange Cyclopean scene. As I was walking along I met two large tortoises, each of which must have weighed at least two hundred pounds: one was eating a piece of cactus, and as I approached, it stared at me and slowly stalked away; the other gave a deep hiss and drew in its head. These huge reptiles surrounded by the black lava, the leafless shrubs, and large cacti, seemed to my fancy like some antediluvian animals.

Like many who have encountered them, Darwin tried to hitch a lift on the tortoises' backs but evidently found it difficult keeping his balance. He estimated that they walked at a pace of 330 metres (360 yards) an hour, equivalent of 6.5 kilometres (4 miles) a day. One would imagine that in keeping with their slow way of life a tortoise's rate of growth would also be slow, but they grow quite rapidly and are said under optimum food conditions to double their weight annually in their first few years. One individual increased from 13 kilograms (29 pounds) to 158 kilograms (350 pounds) in just seven years. All tortoises are vegetarians, the diet of giant tortoises consisting of spiny shrubs, cacti and the fruit of the machineel tree which is deadly to most other animals. Unlike other reptiles, tortoises lack teeth; instead they shred food in their sharp, horny jaws which never wear out.

Darwin spent five weeks on the Galápagos studying the unique wildlife and found that the isolation from land had produced on different islands of the group different species which had adapted themselves to the various conditions. For example on the most arid of the islands he found a race of turtle with a saddleback-shaped shell which he rightly presumed to be an adaptation to the animal's diet – for here the main food consisted of clumps of tall-growing cacti, the shell in this case giving the tortoise the required freedom of movement to stretch out its long neck.

Exploitation

Long before Darwin arrived on Galápagos, the islands provided a convenient stopover point for American whaling vessels en route for the southern oceans. During their stays the crews of these ships slaughtered tortoises for food. They were according to one sailor 'extraordinarily large and fat, and so sweet that no pullet eats more pleasantly'. According to one

biologist who examined the logbooks of 105 American whaling vessels for the years 1811–46, some 15,000 tortoises were killed by the crews of these ships and, since the islands were discovered, around ten million tortoises must have been taken. It is amazing that the tortoises survived at all. The real threat came in 1832, however, when a penal colony was founded on one of the islands. With man came his animals – dogs, pigs and rats ate tortoise eggs and young and the goats which were introduced stripped the islands of their already sparse vegetation. It was not until the early sixties that the tortoises received protection with the establishment of the Charles Darwin Research station financed in part by the World Wildlife Fund. At the station the rarer races of tortoises are reared and fed until they are large enough to be released to restock their native island population.

There is no need to go as far as the Galápagos, however, to research the exploitation of tortoises – of the 300,000 tortoises imported into Britain from North Africa each year less than five per cent survive their first winter.

Above: The giant tortoise's long neck and powerful jaws allow it to tackle most types of vegetation including long-spined cacti.

The House Fly
Nasty, brutish and short-lived scavenger

The housefly (*Musca domestica*) is the shortest-lived insect: males live just seventeen days and females twenty-nine. It is also noted for its filthy habits. It lays its eggs in dung or rotting vegetation and eats either liquid food or liquefies it by regurgitating its saliva. Wherever man lives the housefly is his constant companion, be it in temperate latitudes or the tropics. In Britain the housefly often goes by unnoticed until the middle of the summer when one will land on the table and cautiously approach a piece of food. Its mouthparts are complicated and under the microscope look quite obscene, the large labium forming a trunk or proboscis which when in use becomes turgid by blood being pumped into it from the head. The actual structures that come into contact with the food consist of a pair of soft pads which act like sponges in sucking up the food through numerous very small channels called pseudotracheae.

Its indiscriminate feeding habits cause the fly to be a serious transmitter of disease especially in the sub-tropics where there is poor hygiene. The common housefly is known to be a vector of intestinal disorders which lead to a high infant mortality rate as well as more dangerous diseases such as dysentery, typhus and poliomyelitis. Although we in Britain are now protected from flies by carefully packaged food and refrigerators and the regular removal of domestic waste, one hundred years ago the situation was quite different and in those days horse-drawn transport and the piles of horse dung littering the roads provided this fly with an abundance of sites in which to lay its eggs. Whereas other flies concentrate on cow dung, the housefly seems to prefer human faeces and horse dung but if these are absent it will lay its eggs in well rotted garden compost. The eggs are white, elongate and about 1 millimetre (0.04 inches) long and are laid in masses of 100–150 at a time, each female being capable of laying up to six batches. Less than twenty-four hours after the eggs are laid a tiny cream-coloured maggot emerges which moves by wriggling its way through its massed food by alternate contractions of its body. Its body is soft and vulnerable, the only hardened parts being two mouth hooks which are used as anchorage to haul the animal along and to scratch and tear up the food which is taken like the adult in liquid or semi-liquid form. Within the short space of four days the maggot moults three times before withdrawing to a cooler part of the dung heap where it transforms itself into a dark brown pupa. Three days later the adult hatches out.

The mating behaviour of the common housefly was first recorded over 200 years ago by a French naturalist named Réamur. The male seizes the female in mid-air then making an often forced landing to the ground jumps onto her back and seizes her with his first pair of legs. As a prelude to copulation the male fly gently probes the back of the female's head with his proboscis and lowers his abdomen. The female, in turn, stretches out her abdomen and inserts her ovipositor into the male's reproductive opening.

An American entomologist, L. Howard, estimated that a female fly that began egg-laying in the spring is capable of producing a progeny of 5,600 billion flies by the early autumn, such is its remarkable power of reproduction. We may be bothered by the odd persistent individual but in the tropics where there are optimum conditions of heat and food, whole walls may be black with thousands of bodies of adult flies.

Although notorious as a carrier of disease the fly itself is meticulous in its own cleaning behaviour. Evidently the fly goes through a daily routine of several cleaning motions tackling each part of the body in turn using its saliva and the thick brush of hairs on its forelegs. On the end of each of its walking legs the fly has a pair of soft pads called pulvilli and a pair of claws which allow the fly to walk up sheer vertical surfaces and even upside down on ceilings. It is not known if adhesion to the surface is by suction alone or by secretion.

A fly approaches the ceiling at a low angle with the forelegs extended forwards and upwards. As soon as these make contact the fly somersaults round coming to rest with its head in the direction from which it came.

Right: The housefly (*Musca domestica*) is a short-lived insect with a nasty reputation – known to transmit over thirty viral and bacterial diseases by feeding indiscriminately on animal excrement and human food.

Where do flies go in the winter? Like many other insects including butterflies, the housefly was thought to hibernate in the shelter of a barn or house, but it has now been proved that a small number of houseflies are capable of breeding at a reduced rate throughout the winter. Larval and pupal stages have also been found in exposed places and it is probable that breeding does occur all through the year, the fly having an ability to seek out the warmest and most sheltered micro-environments in any given area. The majority of flies, however, are killed by cold weather and before this a large number succumb to a horrible death – fungal spores that land on their bodies eventually develop into an extensive network within their bodies, erupting through the skin to kill them.

Other Fly Species

The next most common indoor fly is the lesser housefly (*Fannia canicularis*) the males of which are infuriating in the way they will endlessly circle in the middle of the room chasing each other for hours on end. The lesser housefly has slightly different breeding habits, preferring to lay its eggs in urine-soaked material, a pile of unwashed babies' nappies making an ideal breeding ground.

The other large common flies include the bluebottle (*Calliophora erythocephala*) and the greenbottle (*Lucilia sericata*). The bluebottle comes into the house in search of a nest site, it only takes purely liquid food and soon tries to escape though it is exasperatingly difficult to remove, often buzzing around almost at random taking a long time to find an open door or window. The bluebottle mainly lays its eggs on fresh animal corpses which it detects by its keen sense of smell. The larvae on hatching first tackle the fluid between the muscles but as they grow in size the alkaline contents of their excreta neutralize the acid that accumulates in the muscles after *rigor mortis* sets in and they work their way in a heaving mass through the bulk of the muscle tissues reducing the corpse to mere skin and bones within a few days. Bluebottles will also lay their eggs in cooked meat and cheese although they tend to avoid really putrefying meat.

The greenbottle may sun itself on your doorstep or congregate around open dustbins but it rarely comes inside. This fly chooses offal and remains of fish as places to lay its eggs. Because of this it is particularly numerous around slaughterhouses and fish markets. These flies also cause much damage by laying their eggs in moist parts of a sheep's fleece particularly around the anus and underparts. The maggots at first work their way through the wool but as they grow they attack the flesh causing painful sores. In the tropics the habits of flies are truly ghastly: some lay their eggs only in open sores, some in the moist corner of the eye while the maggots of certain species particularly suck the blood of native peoples.

The Robin

Red for danger!

You may assume that the robin you see hopping about your garden in successive years is the same bird, but in fact the robin, with the swallow, is the shortest-lived of British birds, and so every year it is a new bird of the young generation. The robin (*Erithacus rubecula*) certainly packs a lot of activity into its short life of just thirteen months. Though depicted as a tame bird of gentle disposition, it is in fact the most aggressive of our garden birds especially in relation to others of its kind. Its most aggressive behaviour, reasonably enough for a bird which only has one season in which to breed, is connected with the defence of its feeding and breeding territories.

The most vigorous threat display is shown early in the spring when another robin attempts to stake out a territory by displacing the present occupier. Once it crosses a well-demarcated boundary line the resident bird flies up to a prominent perch and utters its short liquid song which is immediately answered by the other male bird. The song serves as a warning and if the force of delivery is insufficient to deter the intruding male the resident bird approaches the other, puffing its feathers out to display the maximum area of its red breast and all the time continuing its intensive warbling. This threat behaviour is nearly always successful and the intruder is quickly chased out of the territory. If however the other bird stands its ground a fight ensues with both birds fluttering to the ground pecking savagely at each other's heads. The newcomer generally gives up by this stage but if it persists the pecking gains in intensity, feathers fly and even blood may be drawn. Although rare, robins have been known to kill each other by a sharp peck to the base of the skull. David Lack, the British expert on robin behaviour, found that the robin reacts violently even to a set of red breast feathers placed within its territory.

Breeding

The problem for the aggressive male robin is enticing a female into his territory; the male at first treats her like another rival and there is some fighting, but eventually he accepts her and they set up a joint breeding territory. In England robins pair from late December onwards and although the eggs are generally not laid until March or April there have been records of robins breeding as early as January. It is the hen alone that builds the nest, a rather bulky structure made of moss and dead leaves lined with hair. The nest is generally well-hidden in a hole in a bank or a tree stump or amongst dense vegetation. In gardens lacking in natural cover the robin is renowned for its choice of unusual nest sites – old kettles, tin cans, even in the pocket of an old jacket. Nest building is completed in a very short time, an average of four days, but can be a lot quicker as witnessed by one Hampshire gardener who hung up his coat in a potting shed in the morning and returned at mid-day to find an almost complete robin's nest in one of his pockets.

From the time of nest building until the young are incubated the male robin feeds the female with a constant supply of insect food, mainly small caterpillars. The normal clutch consists of five white eggs speckled with reddish-brown and there is generally a second brood in late April or early May. Once established in a breeding territory the female bird (which also has a red breast) defends the territory against other robins both male and female. Both parents feed the speckled young and help protect them against predators by alarm calls. If a cat or weasel approaches the nest the parent birds perch well out of range and give a sharp scolding 'tic-tic-tick' which causes the young to crouch down and flatten themselves against the floor of the nest, their brown plumage serving as perfect camouflage. If a hawk is spotted flying overhead a parent will 'freeze' and utter a high-pitched 'see-eep'.

By July breeding is over, the parent birds moult and the young acquire the red breasts of the adult bird. This is the only period of the year when the robin stops singing and by the end of August both male and female birds have set up feeding territories and begin again their short warbling song. Male birds and some females remain in their territory throughout

Right: During its short life of just thirteen months a robin (*Erithacus rubecula*) may rear two or even three broods.

the winter months but most females and all juvenile birds disperse to more favourable winter feeding stations.

Relations with Man

The habit of a robin following the gardener around and perching on his spade is merely an extension of the robin's wily tactic of shadowing larger birds and mammals that disturb insects on which the robin feeds. Robins often accompany pheasants, for example, who when scratching about on the ground disturb insects which are quickly pounced on by the more nimble robin. Even moles are followed and a robin has been seen to snatch an earthworm almost from the mouth of a mole. In England the robin's tameness towards man is probably due to a lack of persecution and even a tendency towards encouragement. In contrast robins on the continent are trapped and netted for food, and so are shy and retiring and generally found in thickly wooded and forested areas. On the continent the robin is replaced as a tame bird by the related redstart (*Phoenicurus phoenicurus*) in low-lying country and by the black redstart (*P. ochruros*) in mountain areas.

There are an estimated five million pairs of robins breeding each year in Britain and in addition to their familiar habitats of garden and woodland robins have successfully colonized the dense stands of conifers in Forestry Commission plantations.

The Axolotl

Ageless amphibian

One of the strangest of life histories and one which still perplexes zoologists is the case of the axolotl, the larval form of the Mexican salamander (*Ambystoma mexicanum*). The exotic name of this creature means water-monstrosity or water-dog and the animal is said to represent the Aztec god Xolotl who resided over the souls of the dead. This lugubrious association is readily understood when one views the animal, for the axolotl is like a giant newt larva, up to 30 centimetres (1 foot) long, naturally dark coloured, but bred in aquariums as translucent white with bunches of salmon-pink gills on each side of the head.

Although its life was well known to the Mexicans and recorded by a Spanish writer as early as 1628, it was not until the mid-nineteenth century that live specimens were seen in Europe, a handful of animals being shipped from Mexico to the Natural History Museum in Paris. Evidently, they thrived and their progeny were sent all over the world to be viewed by the curious. The axolotl's strangest feature is that it is neotenous – it may retain its juvenile characteristics throughout its life, and actually breed in the larval state, a phenomenon known as paedogenesis. A few years after the first shipment of axolotls, a German naturalist, Marie von Chauvin, discovered that when her larvae were deprived of water,

Below: The axolotl changes into the adult form (illustrated) when the lakes in which it lives dry out.

they metamorphosed into adult salamanders. This is exactly what happens in the wild state for, normally, the axolotl lives in shallow freshwater lakes which are subject to drying out. Axolotls respond to this environmental stress by transforming into the better-equipped adult form, a sturdy, broadheaded salamander with vertical grooves down its sides which mark the position of the ribs. Professor Garstang, an eminent zoologist noted for his rhyming couplets to describe the often tortuous transformations of animal larvae, succinctly tells the axolotl's story in his book, *Larval Forms with Other Zoological Verse*:

Ambystoma's a giant newt who rears in swampy waters
As other newts are wont to do, a lot of fleshy daughters:
These Axolotls, having gills, pursue a life aquatic,
But, when they should transform to newts, are naughty and erratic;
They change upon compulsion, if the water grows too foul,
For then they have to use their lungs, and go ashore to prowl:
But when a lake's attractive, nicely aired and full of food,
They cling to youth perpetual, and rear a tadpole brood.
And newts *Perennibranchiate* have gone from bad to worse:
They think aquatic life is bliss, terrestrial a curse.
They do not even contemplate a change to suit the weather
But live as tadpoles, breed as tadpoles, tadpoles altogether.

The axolotl is the subject of numerous research papers mainly dealing with the response of the animal to different levels of thyroxin, the hormone involved with amphibian metamorphosis. And because its body organs easily regenerate, it has been the subject of skin graft techniques and also cancer research. Indiana University even brings out a twice-yearly *Axolotl Newsletter* dealing, amongst other things, with members' problems of how to cope with diseases prevalent among their aquarium-bound pets. Recently, on British television, in a rather serious play, one of the characters kept an axolotl in the confines of his prison cell! Although the axolotl has become something of a cult animal in the advanced countries, in its native land it is viewed rather differently and is considered a choice food item.

Other Neotenous Species

The axolotl is not the only neotenous amphibian; the related tiger salamander (*A. tigrinum*) of North America fails to metamorphose in the colder northern parts of its range, and there are others Garstang mentions that live permanently in the larval state. One of the most curious is the olm (*Proteus sanguinus*) which lives in underwater limestone caves in Yugoslavia. No matter how large a dose of thyroxin is administered, the olm fails to 'grow up', though its skin does darken from its normal ghostly-white colour. This slender amphibian lacks eyes but has patches of light sensitive cells in their place. It is said to lay eggs mainly but has also been known to give birth to live young.

Even more extreme in appearance are members of the family Sirenidae which have permanent larval forms but lack a pelvis and hind limbs and have extremely reduced front ones. They are noted for their slow growth rate and longevity – they have been kept for over fifty years. One of the more common species is the great siren (*Siren lacertina*), found from South Carolina to Florida. When its pools dry up, it simply burrows into the mud where it can remain for many months, protected by mucous.

Two amphibian families show incomplete metamorphosis. These include the Amphiumidae members, which are commonly known as Congo eels although they inhabit the southeastern parts of the United States. These amphibians are particularly nasty, grow up to 90 centimetres (3 feet) and are armed with strong teeth which can inflict a painful bite. The adults lose their gills but retain other larval features such as lack of eyelids. The other family, the Cryptobranchidae, contains giant types which grow up to 115 centimetres ($4\frac{1}{2}$ feet) and may weigh over 10 kilograms (22 pounds). These large sluggish salamanders tend to live in the shady recesses of clear, fast-flowing streams where they feed on fish, frogs, crayfish and insects.

Neoteny is not a new phenomenon, for fossil evidence showed that it existed over 250 million years ago in the Permian period in the case of a salamander by the name of *Dvinosaurus*.

Overleaf: One of nature's oddities – an axolotl, the larval form of the Mexican salamander (*Ambystoma mexicanum*), which can breed without ever changing into the adult form.

The Queen Termite

Aged mother of the insect world

Entombed within their earthen fortress a queen termite and her faithful mate may live for a hundred years or more, at times producing eggs at the phenomenal rate of 8000 – 10,000 per day. These insects are also renowned for their destructive forays on property and buildings. People living in Africa have left their houses for an evening to return to find furniture, books and bookcases all reduced to fragments. Everything wooden is susceptible to attack, including farm gates, carts and huge tree trunks. Most species not only attack at night but work undercover, streaming out in their thousands along specially constructed runways which lead just below the surface of the soil. If a large branch or trunk is encountered the termites cover it with earth then set to work until all that remains after a few days is the bulging impression of its shape.

The great fortresses that dot the African savannah are generally the work of *Macrotermes bellicosus* and may reach 12 metres (40 feet). In Australia's Northern Territory *Nasuitermes triodae* also builds tall steeple-like structures. It is here also one finds the precision-made structures of the compass or meridional termite (*Omitermes meridionalis*) which attain a height of 3 metres (10 feet) and are flattened from side to side in such a manner that the narrow ends face north and south and the broad faces east and west. This regular pattern of building is probably connected in some way with temperature regulation. In the wetter regions of tropical Africa one finds the curious domed nests of *Cubitermes* which serve as an umbrella in deflecting heavy rain from falling on the main part of the colony. But these showy monuments are just the 'tip of the iceberg' for most termites build their nests underground with perhaps a slightly raised mound to indicate the site of the colony.

The hub of any termite society is the nuptial chamber where the queen, who often measures up to 25 centimetres (10 inches), is served by a constant stream of workers who feed her and in return take up drops of liquid that are exuded from her abdomen. Meanwhile other workers groom her monstrous flanks while others work like a chain gang to remove the clusters of eggs as they are laid taking them to special nursery chambers where they undergo their development. The workers also have to feed the soldier termites. Distinguished by their larger size and disproportionately large heads armed with serrated jaws, these soldiers are incapable of feeding themselves and solicit the workers for food by nuzzling their heads and abdomens. In some species a second set of workers lack armed jaws but instead have the front of the head elongated into a tubular trunk with a glandular tip. From this it can squirt viscous fluid capable of entangling and killing predatory insects, notably ants which constitute the main threat to every termite colony.

The workers though constantly busy serving both the royal pair and soldiers do however spend a lot of time passing food between themselves which not only serves a nutritive function, but also helps bind the colony, maintaining a unified cohesive force. One finds a similar situation among a hive of bees where the queen exudes a secretion which is passed from worker to worker.

Below: The fortified earth mounds of *Bellicositermes* dot a savannah landscape denuded of trees by elephants. After the trees have been killed by over-browsing the termites feed on the dry dead wood.

Above: The enormous queen termite receives lavish attention from a mass of swarming workers who feed and groom her and collect, in return, a constant supply of eggs.

To digest their often tough fibrous food most termites contain within their guts minute protozoan animals which break down the cellulose of plant fibres into a form that is more easily assimilated. Those termites that lack this digestive aid probably have their own bacteria which act on cellulose, but others do not have the same requirements as they feed mainly on fungi which are cultivated on moist beds of premasticated food which are laid down in chambers often located near that of the king and the queen.

Forming a Colony

In the topics, after the rainy season, activity inside the colony takes on a dramatic turn. The workers intensify their activity excavating narrow burrows that lead up to the surface and are used as escape channels for numerous young winged males and females that prepare for their once-in-a-lifetime nuptial flight.

Once the 'doors' are opened the winged forms swarm out rising like clouds of smoke into the air; it may take a few hours before all the termites have flown the nest. Although they are guarded at the entrance by the ferocious soldiers, many of them never get off the ground. The aardvark, a mammal that exists almost solely on termites, greedily laps them up with his long sticky tongue, while toads, lizards and ants also feast on this glut of food as well as certain African tribes who find the swarming insects something of a delicacy.

The few that do survive are the founders of new colonies. As soon as they touch the ground the young reproductives run frantic-ally around with wings outspread though the wings soon break off after the first frenzy of activity. Different species have various types of courtship behaviour but it inevitably leads to a couple walking off (male behind female) into the bush to chose a prime nest site. A pair may wander for several days the male keeping his mouth close to the female's abdomen at all times. Eventually a suitable site is found, they dig a hole, construct their nuptial chamber and it is only then that mating takes place. The young couple exist at first on the wing stumps which are reabsorbed into the body and then later on the first batches of eggs that are laid. As the colony expands and a sufficient number of workers reach maturity the couple become dependent on them and are blockaded in. If a colony is overrun by predatory ants or destroyed in some other way it is capable of regeneration even though its original founding members may have been destroyed. Termites known as secondary reproductives take over, the colony probably being aware of its loss by the termination in supply of the queen substance exuded by the queen termite. It is as though by nipping off the terminal shoot the axillary ones develop. In this case there may be more than one functional surrogate queen as they appear not to have the reproductive potential of their 'mother'.

Though not as impressive as some of their tropical counterparts, termites are found in the Mediterranean and there have been isolated reports of them further north. Perhaps in the future these insects may find a way of permanently colonizing more temperate regions.

The Asian Elephant
Wise old man of the forest

Elephas maximus, the Asian elephant, has a lifespan that is just a little longer than the lifespan of a human being. For most of our history people could expect to live, on average, for about thirty-five years. In the wild an Asian elephant lives probably for about fifty years and in captivity in India one elephant is known to have lived for 130 years. African elephants do not live for quite so long.

There is a general rule that large animals live longer than small ones. The reason is fairly obvious. It takes much longer for a large animal to grow to its full size, and so to maturity, than it does for a small animal. It is not true that a large elephant will live longer than a small elephant, but elephants do live longer than mice, dogs, cats, or most other mammals.

At Work

The fact that the lifespans of the elephant and the human are so similar has allowed a partnership to develop between them that is quite unique. There are many people who own animals to which they are devoted, but the relationship usually ends with the death of the animal. Where the partner is an elephant, it is quite likely to be the human who dies first and this means a human and an elephant can meet when both are young and then remain close friends for the rest of the human's life. A working elephant usually has only one human handler, and they spend many years together.

Elephants do not breed well in captivity and so most working elephants begin their lives in the wild. An elephant cow may produce her first calf when she is fifteen to twenty years old and during her lifetime she may produce four or five offspring, seldom more. Twins are uncommon. The gestation period for an elephant is twenty to twenty-two months and very often the birth is assisted by another female – a midwife.

Elephants like company and the herd is a large family group. The baby elephant and its mother travel with the herd and the first introduction the calf has to a new way of life is when some of the herd are lured into an enclosed space from which they cannot escape. Then another elephant, a stranger, will separate some of the weaned calves from the adults. Soon the calf will meet a boy and so commence the lifelong partnership. The adult elephant and the human instructors help boy and calf to learn together, and they establish a regular daily routine. This allows ample time for feeding, for an adult working elephant needs over 200 kilograms (440 pounds) of hay and nearly 300 litres (66 gallons) of water a day, and plenty of time for wallowing in the cool river mud and showering. Asian elephants cannot tolerate great heat.

Working time is most likely to be spent moving, stacking and loading timber. The elephant handles large, heavy objects with great delicacy and precision, but the task does not over-tax its great strength and it is easy to learn. Wild elephants push over trees and handle them with their trunks as a matter of course. They are strict vegetarians, feeding on grass, bamboo and the fruits, leaves, twigs and bark of certain shrubs and trees. When desirable food is out of reach the elephant simply fells the tree, or shakes it to remove fruit. This spoils the habitat for many other species and it causes considerable damage, especially when a herd of elephants invades a tree plantation, but it does prepare the animal for a life shifting timber.

In the past elephants have enjoyed more glamorous professions. They have borne emperors and have carried warriors into battle. Miniature hunting lodges, called 'howdahs', have been built on their backs for expeditions into the jungles. In Hindu legend, where the elephant-faced god Ganesha is the divinity who clears away obstacles, it is believed they once possessed the power of flight. It was taken from them by a hermit who lived among the roots of a banyan tree after some elephants had landed on his tree and demolished it.

In the Wild

Elephants like to be close to other elephants. They are forest animals, except where they have demolished the forest and created grassland, and when individuals are out of sight of one another because of the trees, they make a

low rumbling sound that keeps them in touch. Should danger threaten the rumbling stops and the elephants are alerted.

They are gentle animals, with good memories, though perhaps not so good as legend would suggest, and they are very intelligent. They have few natural enemies, do not hunt, and so have no need to be aggressive. Occasionally, however, an old bull may become mentally unstable and be turned out of the herd for its antisocial behaviour.

An adult bull Asian elephant stands about 3 metres (10 feet) tall at the shoulder and weighs about 6 tonnes. Only the males grow tusks, and these are usually about 1.5 metres (5 feet) long and weigh about 16 kilograms (35 pounds) each, though tusks of twice that length and more than twice that weight have been known. Not all males grow tusks, however. There are four distinct races of Asian elephants – which are all quite different from the two African species of elephants – from India, Sri Lanka, Malaysia and Sumatera. The Sri Lankans seldom possess tusks, and the Sumaterans are rather smaller than the other races and have longer trunks.

Above: Mother and calf Asian elephants (*Elephas maximus*) – apart from man, the longest-lived land mammal.

The Dragonfly
Insect of antiquity

Above right: An *Aeshna* dragonfly shows the supreme mastery of flight exhibited by this ancient order of insects.

Below right: Damselflies are distinguished by their delicate paddle-shaped wings which are folded behind the body when at rest. Here a male *Coneagrion puella* is shown devouring a mosquito.

This ancient group of insects has existed relatively unchanged for 260 million years. In Carboniferous times giants of their kind with a wingspan of over 60 centimetres (2 feet) glided over a landscape very different from the one we know today. Along the antediluvian swamps flanked with giant ferns and horsetails these dragonflies hunted primitive mayflies and cockroaches plus many other insects that are now long extinct. The largest of the extinct species was *Meganeura monyi* which had a wingspan of 70 centimetres (27 inches) and whose larvae must have been at least 30 centimetres (12 inches) long.

Today there are no giants, the largest species having wingspans of less than 20 centimetres (8 inches), but there do exist a few representatives of a family, the Petaluridae, which flourished some 150 million years ago in Jurassic times. Of the nine known petalurid dragonflies over half live in Australia and New Zealand. Though not the largest of species they are extremely bulky insects distinguished from other dragonflies by their heavily veined wings, the stout cylindrical abdomen and the large leaf-like appendages located at the end of the abdomen of the males used to grasp the female. The larvae are also curious in that whereas most others live in still, open water, these types burrow in the mud in spring-fed bogs and in swamps along the sides of streams. This habitat is thought to be similar to that of long ago. The burrows though opening onto 'dry' land are up to 1 metre (40 inches) deep and lie below the water table. Hunting mainly at night petalurid larvae feed mainly on earthworms and live in the larval stage for five to six years. The bulkiest of these types is *Petalura ingentissima* which has a wingspan of over 15 centimetres (6 inches) and feeds mainly on other kinds of dragonflies caught in flight.

Dragonflies are the swiftest of flying insects, and all species are carnivorous, feeding on other insects which are caught on the wing. The prey is detected by mosaic vision made up of the sum of images received by 1000 or more individual 'cameras' that make up the two compound eyes. The prey is grasped by the strong legs which are held in front of the body to form a type of basket. Dragonflies like sunny climates and of the 5000 known species most live in the tropics and the Mediterranean region. In Britain there are some forty species.

The two main groups of dragonflies can be distinguished by the way the wings are held when resting; in the more heavily built members of the Anisoptera the wings are extended by the sides whereas in the more slender Zygoptera, the damselflies, the wings are folded together over the back.

Breeding and Feeding

A curious feature of dragonflies is their unique method of mating. They can often be seen during the summer flying in tandem with the male in front holding the female by the neck with clasping organs on the tip of his abdomen. The actual copulatory organs are not as in most insects situated at the end of the abdomen, but in a pouch in the anterior part. Before seizing the female the male has already discharged sperm into this pouch; and during flight the female curls the tip of her abdomen into the pouch to receive the sperm.

The eggs are laid in water or on weed just above the surface. The larva that hatches is one of the most predatory of all insect types, and will tackle animals much larger than itself including tadpoles and young fish. It generally stalks its prey slowly and cautiously along the bottom of a pond or between dense masses of aquatic vegetation, but, when within striking range, will suddenly move forward at lightning speed by a type of jet propulsion whereby water is forcibly ejected from the end of the abdomen. To seize the prey the larva is equipped with a structure known as the 'mask', a development of the lower lip or labium. When not in use the mask is folded under the head but can suddenly be shot out at the prey which is secured by a pair of sharp hooks at the tip of the labium. The mask is then retracted and the prey is chewed by the mandibles.

In Britain the most brilliantly coloured species is the damselfly *Agrion splendens*. The male is metallic ultramarine-blue.

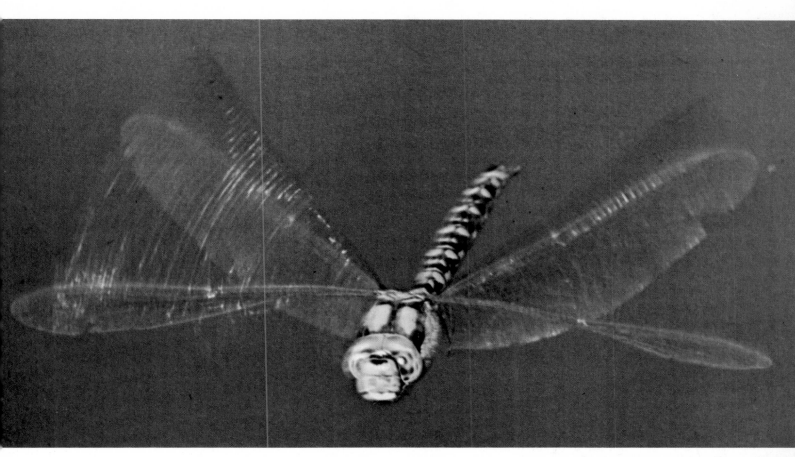

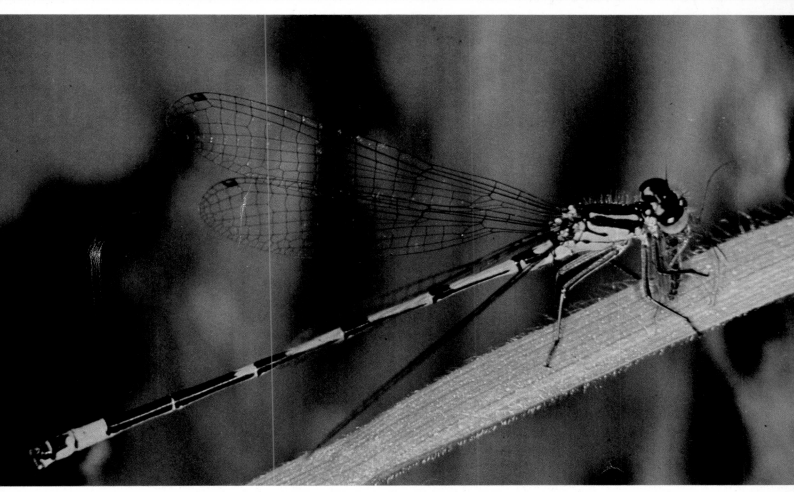

SHORT AND LONG LIFE-CYCLES

Name	Hatching of eggs	Larval or foetal life
Man *Homo sapiens*		266 days Foetus in womb
Chimpanzee *Pan troglodytes*		225 days Foetus in womb
Asian Elephant *Elephas maximus*		607–641 days Foetus in womb
Common shrew *Sorex araneus*		20 days Foetus in womb
Duck-billed Platypus *Ornithorhynchus anatinus*	12 days Incubated by female on nest of leaves.	
Ostrich *Struthio camelus*	42 days Clutch of 15–20 eggs buried in sand and incubated when the air is cold by male and female.	
Giant Galapagos Tortoise *Geochelones elephantopus*	4–8 months Eggs are buried and left by mother.	
Alligator *Alligator mississippiensis*	2–3 months Buried and guarded carefully by mother	
Axolotl *Ambystoma mexicanum*	8–9 days Can vary considerably depending on weather.	
Common frog *Rana temporaria*	14 days Laid in mass of jelly in water.	12 weeks As aquatic water-breathing tadpoles.
Dragonfly *Pyrrhosoma nymphula*	18 days Deposited on water plants by female dipping the tip of her abdomen under the water.	2 years as aquatic nymph
House Fly *Musca domestica*	8–48 hours Eggs laid on dung of large mammals Development depends on temperature.	6–18 days in 3 larval instars 5 days as pupa

Infant and juvenile life	Adult life
Lactation: very variable, 2 months–2 years Infant: 6 years Reproductive age: 15 years	70–75 years
Lactation: 2–3 years Infant: 3 years Reproductive age: 6–10 female 7–8 male	35–40 years
Lactation: Up to 2 years Reproductive age: 16 years female 14 years male	Over 70 years
Lactation: 21 days Reproductive age: 7–10 months	Average lifespan 11 months
Lactation: 4 months Reproductive age: 18 months	Not known. A minimum of 5 years is probably usual.
Fledging: 6 months Reproductive age: 4 years	Over 50 years in captivity. Many die in accidents, often from broken legs.
Reproductive capability depends on size and therefore food supply. Varies between 6 and 35 years.	100 years in the wild. 150 years in captivity
Reproductive age: 6 years	Over 50 years
	Over 25 years
Reproductive age: 3 years	12 years has been recorded in captivity
Reproductive age: 14 days	35–40 days
	17 days male 29 days female

THE NUMEROUS
—AND—
THE RARE

A family group of ring-tailed
lemurs *(Lemur catta)*. Lemurs, confined to the
Malagasy Republic (Madagascar), are threatened
because of the destruction of
their forest habitat.

Mice and Rats
The rodent plague

Rats

While *Homo sapiens* has been busy conquering the world, the rats have crept close to him. Where men are, there are usually rats. Their proximity has not been one enjoyed by man, however valuable it has proved to be to rats. Their presence has changed the face of human history more than once. Only a few of the 570 or so forms of rats have sought out the dwellings of men as their preferred habitat, but these few have assimilated new habits with a degree of adaptability that is shown by no other mammalian. They are not highly specialized animals so they are able to live satisfactorily in a wide range of environments, including underground burrows, treetops, houses, dry scrubland, tropical forests, windswept moorlands and inundated river estuaries.

Before adopting the settlements of men, rats lived in large areas around the Mediterranean Sea and the Middle East, south-eastern Asia, India, China, Japan, Australia and Papua-New Guinea. Following man, they soon achieved a nearly world-wide distribution. The brown rat (*Rattus norvegicus*), black rat (*R. rattus*) and the little rat (*R. exulans*) are the main species to bind their destiny to that of man, following him on land and over the seas in ships. Their bones lie with those of early Stone Age people, and the living animals scurry about the great cities of the modern world: Calcutta and New York City include localities where rats are in command, sharing houses, tenements, sewers, warehouses and shops, even hospitals, with their human hosts.

By the sixth century AD, rats had commenced their role as plague bearers: they may have begun much earlier, but history is silent on the matter. The first of three pandemics ran through Arabia and Egypt before striking into the heart of Justinian's Roman Empire, probably weakening it fatally, and on northwards through Europe to Britain and Ireland. Rats carry many diseases: they infect humans and other animals through their bites and droppings, transmitting such diseases as rabies, trichinosis, tularaemia and typhus. They also carry the flea, *Xenopsylla cheopsis*, which in turn bears the bacillus *Pasteurella pestis*, the infection of bubonic and pneumonic plagues.

The second and most terrible of all the known pandemics began in the 1330s and reached its height in Europe during the following decades, after leaving unknown numbers of dead settlements and cities in Asia. In Europe, a horrified population saw rats dying in their streets before the sickness they called the Black Death struck them too. The rat-borne plague ebbed and flowed across the continents for three centuries, producing a final surge in the Great Plague of London in 1665. A third pandemic spread outwards from Yunnan in 1892 and killed between its inception and the end of the Second World War eleven million people in India alone, and the ripples of the outbreak are still observed in parts of South America and elsewhere.

No one knows for certain when the brown rat attached itself to man's environment. There are claims that after earthquakes in 1727 in the heart of Russia, rats were seen crossing the Volga River in thousands to settle in western Europe, but it seems fairly clear that the brown rat was already known in Europe much earlier than that. The brown rat is a strongly-built rodent with short ears and a short tail. It chooses to live on the ground and under it, burrowing tunnels into compost and rubbish heaps, making homes in sewers, houses, animal stalls and storage houses. They are semi-aquatic creatures so they readily adapt to life beside canals and rivers, where they are accomplished hunters of amphibians and water birds. Fishermen have watched rats at dusk and dawn floating in a current and making movements with their forefeet to filter small creatures from the water as it passes: a motion that is called 'the searching grasp'.

The increase in the population of rats in a suitable environment can be rapid. In a bird sanctuary on the island of Nooderoog, in 1945, about 15,000 brown rats in an area of only 15 hectares (37 acres) virtually wiped out the bird-life. They ate their eggs, and caught even sea-gulls. The female brown rat has a heat that lasts only about six hours, but during that time she

Right: The black rat (*Rattus rattus*) is a prolific breeder, capable of six litters a year each of which may comprise ten young. This rodent was responsible for the spread of bubonic plague through Europe in the fourteenth century which claimed twenty-five million lives.

may copulate up to 500 times with various males. Her gestation period of 22–24 days is followed by the birth of about eight young, who are born blind and naked. Several rats may care for the young, motherhood not being an exclusive function in rats, and the young open their eyes after fifteen days or so, by which time they will have grown fur. After another week, they are ready to leave the nest.

A brown rat may produce as many as five litters a year, more in captivity. The young rats mature fast: males are ready to mate at three months and females soon after, so a female may produce hundreds of descendants in one year. Their practice of communal motherhood assists in making the survival rate of the young unusually high. If a mother is killed, her role is continued by other females until the litter is strong enough to fend for itself.

A pack of rats is essentially a large family. Where strange rats are assembled in a place artificially, the males will establish their territory and fight any interlopers to the death, giving vicious bites that bleed the loser fatally. Once he has killed all opposition, he begins to build his pack/family by choosing a mate. She will join him in attacking any rats that enter their territory. In the wild, packs of over sixty rodents are common, and a few include more than 200 animals, who probably recognize each other by smell as well as by sight. Rats are strongly scent orientated animals, their sense of smell being their primary one.

The black rat or house rat has a longer, more slender shape than the brown rat. Its tail is longer than its body and its ears, too, are larger than the brown rat's. Their population is smaller than it was. It is a good climber, and lives in attics and on beams of warehouses and barns. In the days of wooden ships, black rats – sometimes called ship rats – infested the sailors' quarters and the holds. Probably ninety per cent of rats on board ships are black rats. It was these rodents that carried the plague fleas across the seas to bring the Black Death to Europe in the Middle Ages, one of the great catastrophes in human history.

Today, they are a particular danger to people living in the tropics, where they eat human foods in amounts that are unequalled by any other mammalian and by few insects. Rat droppings and urine contaminate huge stocks of stored foods, carrying typhus and several other serious diseases.

Right: The larger, commoner and more aggressive brown rat (*Rattus norvegicus*) is a cosmopolitan pest of town and country. Recent mild winters in Britain have allowed it to breed and become especially numerous. In many areas it has become resistant to powerful poisons.

Mice

Mice, like rats, are among the most fertile and adaptable animals in the world. The distinction between the two is mostly one of size: mice have bodies of less than 15 centimetres (6 inches) in length while rats have longer ones. The house mice have adopted human settlements as their homes, and helped themselves to human provender probably since the appearance of man on the Earth. The two main species to have joined the humans are the western house mouse (*Mus musculus domesticus*) and the northern house mouse (*M. m. musculus*): and of the two, the western house mouse is the one more committed to its human host.

Since early history the mouse has held a place of reluctant affection among most people. The Ancient Greeks used to consult it as an oracle and considered it a symbol of loving tenderness. Its prolific nature encouraged the Ancient Egyptians in the belief that mice were generated by the sun from the Nile mud. Even the philosophic Aristotle was led to assert that they were born from the rubbish in houses and ships. Remarkably, despite their fertility, the females are capable of a form of birth control.

When their population begins to overreach the available supplies of food, their ovaries cease to function and many of the adolescent females become infertile.

House mice are adaptable to wide ranges of temperature. They have even nested in carcases of frozen meat in cold stores, where they successfully continue to breed and raise their large families in temperatures of a mere 10° Centigrade. Their endearing appearance hides hides a deadly danger of disease. Like the rats, their droppings and urine carry the threat of *Salmonella* poisoning and spotted fever, and their parasites can infect with bubonic plague.

A house mouse lives and scavenges in a small area, often no more than the corner of a room. They live, like the brown rats, in family groups, depending on smell and hearing for communication.

Rats and mice have brought starvation and disease to millions of people over the centuries. In return, albino brown rats and house mice now contribute to the prevention and cure of disease in the world's laboratories. While it is not a voluntary gesture to medical science, it is one that humans should have the grace to keep in mind.

The Monkey-eating Eagle

Crested prince of the jungle

Far right: The monkey-eating eagle (*Pithecophaga jefferyi*) raises its head crest when alarmed and also to startle its prey.

One of the rarest sights in the world is to see a monkey-eating eagle soaring over the dense mountain forests of the Philippines. This magnificent bird, over 90 centimetres (3 feet) long with 3-metre (10-foot) wingspan, is the second largest eagle in the world next to the harpy eagle of South America with which it shares its jungle habitat. Like the harpy this eagle also possesses a crown of feathers which are raised when the animal is excited or alarmed – the lance-shaped feathers of the crown when fully extended form a semi-circle setting off the dark feathers of the face, the huge axe-shaped beak and the piercing blue eyes. The whole effect is terrifying and presumably contributes to the shock and confusion of the prey as the eagle makes its final strike.

Although today one of the world's rarest birds, the monkey-eating eagle has only been known in the West for less than a hundred years. In was first discovered in 1896 by the British animal collector John Whitehead who in that year brought a male specimen back to London, and at the thirty-ninth meeting of the British Ornithologists' Club held at the Frascati restaurant in London, described the genus *Pithecophaga* (meaning monkey-eating) and the species *jefferyi* in honour of his father who had financed a number of Whitehead's long-distance expeditions. Since its discovery the eagle has been more fully described and its behaviour observed although the inhospitable and remote nature of its forest habitat has prevented any very detailed studies. We do know that it was quite common on at least four islands of the 7000 that make up the Philippines, but today there are less than a hundred of these birds in remote pockets of true jungle on the islands of Mindanao and Leyte.

The most accurate information on the behaviour of monkey-eating eagles in recent years has been made by the Filipino scientist Professor Gonzales who observed a pair in the southern province of Davao del Sur on Mindanao. The nest was situated in the crown of a tree at least 30 metres (100 feet) above the ground, and without the presence of the adult birds it would have been impossible to locate as it was surrounded by creepers, epiphytic ferns and orchids.

The most remarkable finding by Gonzales was that the very name monkey-eating eagle was something of a misnomer as the pair he observed fed mainly on flying lemurs which accounted for ninety per cent of the prey brought back to the nest. Flying lemurs seem unlikely prey as they are nocturnal but the eagles were observed hunting them early in the morning, snatching them in mid-air as they glided to their roosting sites in the hollows of trees. Stragglers were also picked up as they clung to the limbs of trees, even when they clung to trunks only a metre or so from the ground. But monkeys were also taken, notably the Philippine macaque which roams the forest in family groups led by a large male. Once an eagle has been spotted the male gives a warning cry and the troop dives for cover, often leaving the male exposed to the eagle; but a large male macaque is a formidable adversary and to catch a monkey the eagles tend to work in pairs, either picking off a juvenile or making repeated attacks from all directions on the isolated male. Apart from large prey the eagles were seen to take flying squirrels and a male hornbill which was presumably taken by surprise while feeding the female hornbill at the nest hole. However, a group of hornbills will harry an eagle just as an owl can be mobbed by a group of smaller birds. The rufous and writhed-billed hornbills are also a threat to the eagles' eggs and developing young and are soon driven off if they appear in the vicinity of the nest.

Struggle for Survival

Despite its magnificent appearance and prowess as a hunter the monkey-eating eagle is endangered and the chances of its survival beyond the end of this century are slight unless stringent conservation measures are enforced. The eagle has suffered not only from the normal persecution any bird of prey is subjected to but also from the loss of its forest habitat.

In the Philippines there are numerous stories relating to the strength and ferocity of the eagle – how it has carried a hunter himself carrying a deer back to its eyrie, and there are many reports of children being taken from the very doorsteps of their homes. Of course none of these stories is true as the bird is incapable of lifting such a weight but villagers have reported seeing eagles take chickens, small dogs and even young pigs and because of these attacks the eagle has been shot at whenever it has been seen flying near human dwellings. As it has become progressively rarer in recent years it has also been shot as a trophy. Indeed, a stuffed monkey-eating eagle has become something of a status symbol in Filipino homes and high prices are paid for good specimens. Zoos have also contributed to their decline, for not only are these birds a star attraction but in the interests of conservation many zoos have obtained birds with a view to captive breeding. It should be stated that up until now any attempts at breeding have proved unsuccessful for in the wild these large eagles require a vast breeding territory of up to 10,000 hectares (nearly 25,000 acres). Courtship in the wild also involves an acrobatic high altitude flight with the sexes actually mating in the air. These conditions of course are impossible to simulate in the artificial confines of a zoo – the only hope of breeding comes from artificial insemination which in the long term may prove possible. Attempts at reintroducing birds captured in the wild and fed in captivity have also proved useless as these individuals have lost the instinct to hunt any prey more mobile than a domestic chicken.

Apart from the severe threats to the eagle from shooting and capture, far worse is the destruction of the eagle's forest habitat. It has been estimated that the forest is cleared for timber at the astonishing rate of 170,000 hectares (420,000 acres) a year, one of the fastest rates of forest destruction in the world. Even though Filipino schoolchildren over the age of ten have been ordered to plant a tree every year for the next five years, this will have no effect on the plight of the monkey-eating eagle. Some headway has been made in protection of known eyries but the forest guards appear to be accepting bribes from unscrupulous logging companies. Once timber is felled within an eagle's territory the birds retreat and this situation will continue until there is no forest left. For conservationists the monkey-eating eagle is a splendid example of the richness of the wildlife of tropical forests. To save the eagle means saving the forest and the national heritage of the country, a message that needs reiterating to the Philippine nation before the islands are irreversibly spoiled and lose in the process one of the world's most magnificent birds of prey.

Right: Vanquished predator – the monkey-eating eagle is now one of the rarest birds of prey due to the loss of its forest habitat and because it is shot as a trophy.

OTHER ENDANGERED EAGLES

Spanish imperial eagle
(*Aquila heliaca adalberti*)
Confined to the Iberian Peninsula where a maximum of 80 pairs are known to be distributed mainly in west, central and southern Spain and in the Coto de Doñana National Park in Andalucia.

Madagascar serpent eagle
(*Eutriorchis astur*)
Last seen in 1930. If it still exists, is restricted to the humid eastern forests of the Malagasy Republic.

Madagascar sea eagle
(*Haliaeetus vociferoides*)
Approximately 10 pairs restricted to a small area in the centre of the west coast of the island.

Southern bald eagle
(*Haliaeetus leucocephalus leucocephalus*)
Found in the United States, south of 40°N latitude and in northern Mexico. Fewer than 500 breeding pairs exist, most of these in Florida.

The Desert Locust

A plague of Egypt

When looking at a locust in a laboratory it is hard to share the alarm expressed by people who live in the paths of its migrations. This insect that looks indistinguishable from the harmless, cheerful-sounding grasshopper has haunted the nightmares of men over huge tracts of the world for the whole of recorded history and, doubtless, much longer. The Sumerians of the second millennium BC struggled against plagues of locusts, and the citizens of Nineveh near the beginning of the first millennium BC wrote of the 'noxious locust'. At the same time in China, anti-locust officers gathered information on the movements of the swarms.

There is no way of distinguishing a locust from a large grasshopper by a study of its structure. The differences between the two creatures are that locusts have an instinctive need to gather in groups and to migrate in great numbers in the daytime. Grasshoppers are quite solitary: and, although they will migrate, they do so at night and only in solo flight. It is the character of its behaviour that makes a locust different from a grasshopper, but even this distinction is not wholly satisfactory. Some locusts seem happy to remain solitary, and a few kinds of grasshopper break the rule of their species to migrate in swarms – such as *Melanoplus sanguinipes* in North America. It is, perhaps, best to think of the superfamily Acridoidea of the order Orthoptera, which includes grasshoppers and locusts, as ranging from grasshoppers that never swarm to locusts that do so frequently. In between these extremes are gradations of behaviour, and the insects are named according to their commonest mode. The Acridoidea consists of about 10,000 species of grasshoppers of which only 10–12 species may be called locusts. Two of the most feared species are *Locusta migratoria* which has the greatest range, extending eastwards from Africa to Japan and Australia, and the desert locust (*Schistocerca gregaria*). This is probably the species that plagued the Egyptians when Moses sought to free the Jews from their bondage in the Nile delta. It lives in the vast area between West Africa and India, and Kenya in the south to northern Iran.

Breeding

The desert locust lays its pod (clutch) of sausage-shaped eggs up to 15 centimetres (6 inches) below the ground. A single female deposits four or five pods, each of which consists of 60–70 eggs surrounded by a protective froth. This seals the tunnel from the pod to the surface and helps reduce the eggs' loss of moisture. She lays her eggs in fairly dry soil, but is not too particular about its moisture content or condition, avoiding only waterlogged ground. The egg pods of the species are laid close together, the desert locusts being attracted to each other – and so to one territory – by a pheromone which they detect by touch rather than by smell. In this way, the swarm carries its integrity through to the following generation.

In laying her eggs, the female performs an extraordinary feat in extending her abdomen more than three times its normal length as she digs a hole for the pod. She drives her abdomen into the ground with a rotating movement before laying the pod at the bottom of the hole, but the most remarkable part of the operation is that her internal organs, the alimentary canal, her nervous system and intersegmental muscles stretch to match the extension of the outer segments of her abdomen.

The eggs in their holes absorb moisture from the surrounding soil, doubling their weight. When the conditions are too dry for this to happen, the eggs fail to mature, but they will recover their viability after a shower of rain. Some species, such as *Locustana,* lay eggs that can survive drought conditions, delaying their embryonic development for as long as three years before the process is triggered by rainfall. In ideal conditions, most species of young locusts will hatch in eleven days, and the larva or nymph begins its struggle to reach the surface. It levers itself towards the light by inflating two swellings at the sides of its neck to gain a grip on the soil, contracting its abdomen, finding a purchase point and pushing upwards. The nymphal development of locusts runs

through several moults that occur over a period of about 20–45 days as the insect grows. The first three of these show little change, although the larva's wing pads develop slightly. At the fourth instar or stage, the pads turn upwards and point towards the locust's rear; and at the fifth instar, they grow much larger. In the final instar, the adult emerges and its wings fill with air over a period of about twenty minutes.

Migration and Swarming

Locusts are at their most impressive when they migrate. A great swarm swirls through the warm air, darkening the ground beneath. The migrants rise and fall in the thermals, the edges of their swarm loosening and tightening around the main body. Its speed depends largely on the velocity and direction of the wind and its orientation to the locusts' flight path. They fly at 10–25 kilometres per hour (6–16 m.p.h.) and will battle against a low-speed wind for a few kilometres, but long-distance migrations depend on following winds. These may carry them over great distances. In 1945, a swarm left southern Morocco and in only twenty-four hours reached Portugal, a trip of about 1000 kilometres (620 miles). The swarm is not of consistent density throughout. The locusts at its fringes fly in several directions, some directly at right angles to the general route of the swarm, but after a period of wandering, they turn back towards the main body of their companions. It may be that they are drawn back into the swarm by their need for visual contact with their fellows or they may be attracted by the sound of the swarm and the turbulence of the air that is stirred by millions of wingbeats: no one knows for sure what is the force that holds the swarm together.

The sinister way in which locusts seem to move into landscapes ripe for their foraging led some farmer peoples to adopt the locust as a symbol of divine punishment. The insects are borne on convergent winds towards areas of high rainfall, where the foliage is suitable food and the conditions right for successful breeding. There are several convergent zones, but the one that is most significant to the desert locust is the Intertropical Convergent Zone where the dry continental air of the northern hemisphere and the damp equatorial air meet seasonally. The locust migrations coincide with the occurrence of the zones. In the summer, the desert locust breeds in the ITCZ, which stretches verdantly from West Africa to India. Its offspring return towards their parents' original starting point. Many of these migration routes exist, but they are not followed faithfully: a swarm will sometimes adopt another path to make a longer journey or will develop a series of complex local movements.

The locust's migrations are not easily predictable; their swarms may fail to assemble in

Right: A young desert locust (*Schistocerca gregaria*) – the most destructive insect in the world. Large swarms can consume over 3000 tonnes of vegetation in one day.

some years, and in others the weather conditions produce a pause in the breeding cycle. A downpour of rain in an area which has been stricken by drought may begin the development of millions of dormant eggs to produce a plague of new locusts. One of the great puzzles of the swarming of locusts was solved by the recognition of two phases in the insect's life. Dr B. P. Uvarov, who later became the first director of the Anti-locust Research Centre, discovered that locusts lived through a solitarious phase and a gregarious phase. In the desert locust these phases are marked by obvious changes in appearance as well as of behaviour. A desert locust of the solitarious phase is green or straw coloured to a shade that gives it camouflage among the surrounding vegetation, and its pronotal crest (an arched, saddle-shaped segment that lies behind its head) is well rounded. When the locust reaches the gregarious phase, it turns yellow with a striking pattern of black markings. Its pronotal crest flattens out and even becomes slightly concave. More remarkably, its femur shortens.

The clearly recognizable markings of the gregarious phase may help the locusts to interact with one another during swarming. Locusts in these two phases produce different hatchlings: the gregarious nymphs at their first instar are darker and contain more water and fats than those produced by desert locusts in the solitarious phase. This enables the gregarious insects to live longer without food. Their instinct for associations is much stronger than that of solitarious locusts and they lead more active lives.

When the young locusts walk about in search of food, they tend to gather in warm spots or lush patches of vegetation. As this happens, solitarious insects learn to aggregate or interact: and in conditions where the locusts crowd together in great numbers, subsequent generations will become gregarious locusts ready to swarm. The process can be reversed if a population suffers severe reductions and interaction is reduced. Then gregarious phase locusts will become solitarious.

Fortunately, the breeding areas of the desert locust provide insufficient breeding success for the growth of large swarms. This happens only in years of exceptional rainfall followed by successive seasons of above average rainfall. It is only in these conditions that the locust population can build its numbers over a series of generations to swarm dimensions. These pop-

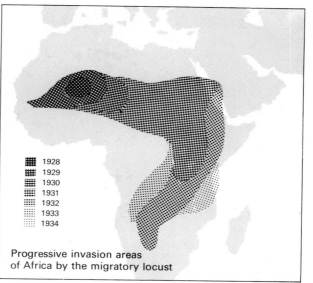

1928
1929
1930
1931
1932
1933
1934

Progressive invasion areas
of Africa by the migratory locust

Left: During the 1930s swarms of migratory locusts, originating in the flood plains of the River Niger, moved south over Africa invading over 17 million square kilometres (6 million square miles) of land.

ulation surges have occurred in the desert locust five times in the last fifty-five years.

The effect of these plague years is stunning. A locust will eat its own body weight of food in one day, that is about 1.5 grams (0.05 ounces) of vegetation. While this sounds little enough, the scale of the damage becomes more daunting when the density of the adult swarm is reckoned at 30–150 locusts per square metre, which amounts to about $45 - 225g/m^2/day$. In a swarm that covers 10 square kilometres these locusts would eat about 2000 tonnes of vegetation each day while they are in the area. The seriousness of this damage varies according to the stage of a crop's growth: seedlings defoliated to this degree would mean total loss for the farmer, but some crops when well established will recover from an attack. The settling of a swarm of locusts on a pasture is especially devastating. In tropical regions the density of cows per square kilometre is about fifteen, and each will graze around 12 kilograms (26 pounds) of fodder per day. A swarm of locusts settling on an area of the same size might eat a thousand times as much: this would constitute a blow to the grazing capacity of the pasture that would take years to restore.

The use of persistent pesticides nearly destroyed the desert locust in its breeding grounds in the Horn of Africa in the decade before 1977. Pesticides, such as aldrin, chlordane and heptachlor, destroyed billions of insects, but authorities who recognized the dangers of the chemicals themselves banned their use: instead, they turned to safer, less effective and more expensive ones.

The Arabian Oryx
Wild white cow of sun and crystal

Two hundred years ago the oryx ranged far across the Arabian peninsula and throughout the arid regions of the Middle East. It lived in herds of between six and one hundred animals covering great distances across the gravel deserts in search of food – grasses, sedges and wild watermelons that stay green well into the hot season. The harmonious picture of these antelopes grazing in the soft light of dawn is, however, seen no more, for in October 1972 the last wild Arabian oryx was gunned down by a hunting party, and the species was wiped out from its native lands. Fortunately the dramatic close to this chapter in its life has not spelt the end of the species.

The Arabian oryx (*Oryx leucoryx*) is a medium-sized largely white antelope with strikingly long curved horns and rather shorter legs than its relatives the fringe-eared oryx (*Oryx gazella*) and the scimitar-horned oryx (*Oryx dammah*), both restricted to mainland Africa. The oryx's horns are used in combat between rival males, as a defence against predators and also, together with the sharp hooves,

are used to scrape out shallow depressions under bushes where the animal rests during the hottest part of the day. The oryx is one of the few larger animals at home in real desert.

Of all the animals of Arabia the oryx was the most highly esteemed by Arab hunters. It was believed that he who killed it achieved the animal's sterling qualities of strength, endurance and courage. In some parts of its range, it was held that he who ate its flesh was endowed with superhuman strength and became impervious to the bullets of his enemies. Moreover the horns of the oryx have long been regarded as one of the great natural phallic symbols. It is therefore no wonder that the numerous desert tribes coveted this quarry, regarding it as the supreme test of manhood. It must have been a real test of endurance, after all, to track and hunt down this animal for miles across the desert on camel or horseback. One could say a natural predator-to-prey relationship existed between the bedouin hunter and the oryx, a balance that remained relatively unchanged until well after the First

Right: A small herd of Arabian oryx (*Oryx leucoryx*) taken from animals bred at San Diego and Phoenix Zoo, has now been returned to the Jiddat al Harasis reserve in Oman.

World War, when, with the advent of modern firearms, hunting was made a lot easier. Indeed, by the 1930s hunters had driven the oryx back to its centre of distribution, to the Nafud Desert on the borders of Saudi Arabia and Jordan and to the Rub al Khali (the Empty Quarter) in the south. The slaughter was speeded up further by the introduction of four-wheel-drive vehicles and automatic rifles. Foreigners stationed in the Gulf and associated with the oil industry are among those responsible for the complete massacre of this animal. The last fresh tracks in the Nafud were seen in 1954, leaving the last stronghold as the Jiddat al Harasis in Oman, though stragglers remained up until the late 1950s in the extreme south-western edge of the Empty Quarter.

Saving Grace

Fortunately there were people concerned with the plight of the oryx and determined to save it. Back in the early 1960s a group of conservationists mounted the far-sighted 'Operation Oryx', the plan being to rescue the last remaining wild animals and transfer them to a safer habitat. Planned by the Fauna Preservation Society and with financial backing from the World Wildlife Fund the 'operation' took off in the autumn of 1961 in the form of an expedition led by a Major Ian Grimwood, a man who had considerable knowledge of wildlife conservation from his work as Chief Game Warden in Kenya. The expedition went into the remotest part of the Aden Protectorate (now South Yemen) and captured three oryx, two males and one female, which were first quarantined at Isiolo in northern Kenya and then transferred to the Phoenix Zoo in Arizona, a location where conditions approximate to those of their native habitat. This small group was to form the nucleus of the World Herd. Zoological Societies and private zoos throughout the world inspired by the rescue operation were eager to help build up numbers. The Zoological Society of London sent a female out to Phoenix where it was soon joined by another female presented by the Ruler of Kuwait. Later, King Saud of Saudi Arabia sent out four males and five females from the Riyadh Zoo, but already before his contingent arrived the first birth in the World Herd had taken place on 26 October 1963. It appeared that the elusive oryx was a relatively easy subject to breed in captivity.

By 1975 there were over one hundred oryx distributed among zoos in Arabia and the United States. The next step – reintroduction back to the wild – was soon to be realized. In February 1978 four male oryx were flown from Phoenix to the Shaumari Wildlife reserve in Jordan. A year later another four animals, this time females, were sent, and this small herd is now breeding successfully.

The other site surveyed for reintroduction is centred on the Jiddat al Harasis, the stony plateau that harboured the last wild oryx. This area, its vegetation and wildlife has been carefully studied for the past few years by Dr Jungius, Director of the World Wildlife Fund's Department of Conservation. This location covers a huge area of 30,000 square kilometres (11,700 square miles) and has uniform geographical features with distinct boundaries which makes it an entity in itself. It also supports an unusually varied vegetation of trees, shrubs, herbs and grasses. But the key to its potential success is the human factor – a small tribe of about 500 bedouin called the Harasis live there. They once hunted the oryx and, regarding it as their tribal property, are in full support of the reintroduction scheme and anxious to prevent further extermination once the animal has been introduced. These people will act as the new guardians of *al m'hat el abiad* – the wild white cow of sun and crystal.

Below: The beisa oryx (*Oryx beisa*) which is confined to East Africa, from Eritrea south to the Tana River in Kenya.

The Quelea
Scourge of the savannah

'Soon after dawn they come, swooping down in their thousands into a field of guinea corn to devour the ripening grain' – a description by ecologist Peter Ward of the most numerous and most destructive bird in the world, the black-faced dioch or quelea (*Quelea quelea*). An estimated ten billion of these birds inhabit the drier parts of Africa south of the Sahara from Mauritania in the west across to Ethiopia and Somaliland. The vast flocks of quelea constitute one of Africa's main agricultural pests affecting the economies of at least twenty-five countries. The quelea belongs to the family Ploceidae or weaver birds, a group well represented in Africa, which also includes the familiar house sparrow. The quelea is a rather inconspicuous, drab-looking bird, and outside of the breeding season both sexes display an overall brownish plumage. Prior to breeding the males show a little flamboyance, developing a distinct black facial mask which extends over the cheeks, chin and throat. Also the bill, legs and a ring of bare skin around the eyes become bright red. In contrast, during the same period the females' legs, bill and eye rims turn bright yellow.

The quelea and related species are often seen in pet shops innocuously bobbing up and down to feed on a spray of millet. In the wild it is an altogether different story where they are feared just as much as locusts for their raids on wheat, rice, millet and guinea corn. Whereas a plague of locusts may attack an area once in five or ten years, quelea constitute a permanent menace.

The quelea may not be the most attractive bird of the African savannah but its behaviour is quite remarkable. It is gregarious throughout the year, feeding and drinking in flocks that may number hundreds of thousands. At dusk these numbers swell enormously, and millions may fly together to a communal roost making a noise like the deafening 'whoosh' of an express train. Although individual birds weigh only around 25 grams (less than an ounce), their combined weight when packed close together in a roost is often sufficient to snap off large branches. The noisy chattering of starlings roosting in the centre of our largest cities is nothing compared with the clamour of queleas settling down for the night in a favoured clump of thorn trees. These roosts may be occupied for only a few weeks before the flock moves on leaving beneath the trees a thick layer of droppings which is collected by Africans because, like guano, it serves as a valuable fertilizer.

Queleas breed during the wet season after rain has been falling for some weeks. By this time there is abundant green grass for nest building, unripe seeds and water for adults and nestlings, and a plentiful supply of insects to meet the extra food requirements of females and young. Colonies are most often founded in tall thorn trees from every branch of which is suspended a neatly woven spherical-shaped nest. Nest building is the work of the male who starts off the construction by making a simple vertical ring of woven grasses. For the quelea this ring also serves as a nuptial chamber into which the female is enticed by the male fluttering his wings in what is known as the 'butterfly posture'. Mating occurs after the display, and once the male is certain the nest will be occupied, he proceeds with building. From the initial ring the nest is extended backwards to make a hemisphere which will eventually form the back of the nest. He then moves to the front of the ring and extends the nest forwards to make the front portion and entrance. The roof area is especially well constructed and is strong enough to protect eggs and young from occasional torrential downpours and the strong midday sun. The nest-building process only takes a few days.

Rapid Breeding

Queleas are remarkably fast and synchronized breeders – it may only take five weeks from the founding of the colony before the young are reared and ready to depart, increasing the huge numbers already in the flocks. All that remains of such a colony are tattered nests, a few feathers and some casualties impaled on thorns; a silent scene, one that a few weeks before was one of constant, frenetic activity.

For a short period during the wet season there is an abundance of wild grass seed which the quelea pick at on the ground, but once the rains cease and the ground dries out quickly under the fierce sun the supply of seeds on the ground is quickly depleted. As the dry season advances queleas concentrate more and more on the fertile alluvial plains, their numbers swelling to form great black clouds that roll steadily over the horizon in search of food. At this time of the year the quelea becomes the scourge of Africa, stripping field after field bare of millet and corn. Damage caused by these birds is phenomenally high, for example in one Nigerian province losses in 1957 amounted to over £1 million. Since the 1950s organizations like FAO (the Food and Agricultural Organization of the United Nations) have tried unsuccessfully to combat the quelea menace. Roosting and breeding sites have been destroyed using explosives, flame-throwing devices and poison sprays. In a small area of Nigeria a control unit destroyed sixty million birds in one season. These numbers, though impressive in themselves, appear to have had little effect on their astronomical population total.

Destruction of birds and nests is anyway difficult. First the colony has to be found and they are often located in inaccessible places such as swamps. Even when location is guaranteed the timing may not be right, for the birds, with their extraordinarily fast breeding rate, may already have flown the nest. Detailed population studies have shown that the quelea population remains fairly constant. Allowing for the fact that on average two young are produced by each mating pair each season, it follows that half of this breeding population, both adults and young, dies every year. Natural predators like goshawks, falcons and bird-eating snakes do little more than harvest the surplus. Man himself, although an unnatural predator, falls into the same category.

The only feasible way of combating the quelea seems to be to grow crops when the birds' wild food is most abundant, a programme which is being realized in many parts of the continent and meeting with some degree of success. It is incredible to believe that in spite of the colossal amount of damage done, the quelea relies on natural grass seeds for over ninety per cent of its annual food requirements!

Above: Throughout the semi-arid parts of Africa flocks of quelea are a scourge as they swoop down on crops in vast numbers.

The Snow Leopard
Ghost of the mountain ridges

In the high mountains of Central Asia between the tree line and areas of permanent snow lives the snow leopard (*Uncia uncia*), one of the rarest and most beautiful members of the cat family. It has rarely been sighted but in 1971 the American George Schaller took the first photographs of it in the wild in the Chitral district of West Pakistan: 'Suddenly I saw the snow leopard. Wisps of cloud moved between us, and she became a ghost creature, appearing and disappearing as if in a dream. We were 120 feet apart on a rugged Pakistani cliff, neither of us moving – two beings bound to each other in a world of swirling snow.'

The kingdom of the snow leopard is indeed remote, vast and inhospitable. It encompasses the mountains and highland steppes of Central Asia from the Hindu Kush of Afghanistan in the west through the Pamir mountains of Pakistan and north-eastwards to the Tien Shan mountains of the Asiatic USSR. Its range also takes in the Himalayan mountain chain across the Tibetan plateau to the Kunlun mountains in the Chinese province of Szechwan.

The snow leopard occupies an intermediate position in the cat family between the great cats which include the true leopards and the smaller cats such as the ocelot. Unlike the larger cats that snarl and roar the snow leopard only makes a purring sound and like a domestic cat tends to eat in a crouched position. One could say that it is a big cat with many of the features and habits of the smaller felids. The snow leopard, or ounce as it is commonly called, is slightly smaller than a leopard, measuring about 2 metres (6 feet 6 inches), with a flattened muzzle, a rather long body and a disproportionately long, well-furred tail. The colour and quality of the fur is however its most distinguishing feature – the ground colour of the coat being a smoky-grey with a yellowish tinge, with black spots arranged as blurred rosettes over the whole body except on the throat and belly. The winter coat is unusually long and has a thick woolly underfur which must be very efficient in insulating the animal from the freezing temperatures. The paws are rounded and have well-developed hairy cushions which act like snow shoes in distributing the animal's weight over a larger area. The pads also protect the paws from rocks heated to high temperatures by the summer sun.

During the summer months the snow leopard lives high up in the mountains. Tracks have been observed and the occasional animal sighted well above 5000 metres (16,500 feet). It is generally nocturnal, emerging at dusk to hunt its favourite prey of ibex, wild sheep and tahr – a goat-like animal with a huge shaggy coat and sharp backward-facing horns. During the breeding season leopards hunt in pairs in order to supply the growing needs of their cubs. They usually take well-worn paths along high ridges scouting the area for prey, then while one rustles up the prey into the valley the other partner waits to attack. The leopard silently pads the ridges and rocky screes stalking its prey until it is within range, then it pounces and clings on tenaciously. In 1958 an Indian zoologist witnessed a snow leopard's attack in the Himalayas. 'I was lying behind a boulder watching the *thar* [*sic*] climbing leisurely up the scree and the rock overhangs towards the north ridge of the Raj Ramba peak when a flash of white and grey fur dived into the spread out herd, and rolled down some hundred feet, all the time hanging on to the young *thar* ewe.' The snow leopard appears to be easily capable of making 15-metre (50-foot) leaps and can scale a slope almost effortlessly with a graceful fluid motion of its limbs.

As winter approaches and the first snows cover the thin alpine pastures, the wild goats and sheep move down into the warmer and more sheltered valleys, and like all large carnivores the snow leopard follows the movements of its prey, lying up in the dense forests of evergreen oaks and rhododendron thickets. In these lower regions the snow leopard also takes prey such as the monal pheasant, deer and wild boar.

Fight for Survival

In severe winters when there is a scarcity of natural prey the leopard will attack domestic

livestock and, apart from nightly raids on sheep and goats, dogs have been taken and animals as large as yaks have been attacked. It is no wonder therefore that it has always been hunted by the local peoples – but the main reason for the decline in numbers of snow leopards has been hunting for its fur. Its luxurious pelt has always fetched a high price on the market, the subtle coloration and rich-ness of its fur appealing to the 'sophisticated'. To quote one advertisement from an American literary magazine in the sixties: 'Untamed . . . the Snow Leopard, Provocatively dangerous. A mankiller. Born free in the wild whiteness of the high Himalayas only to be snared as part of the captivating new fur collection . . . styled and shaped in a one-of-a-kindness to bring out the animal instinct in you.'

Right: The snow leopard's broad paws make it adept at climbing and are ideal for padding over snow-covered mountain ridges.

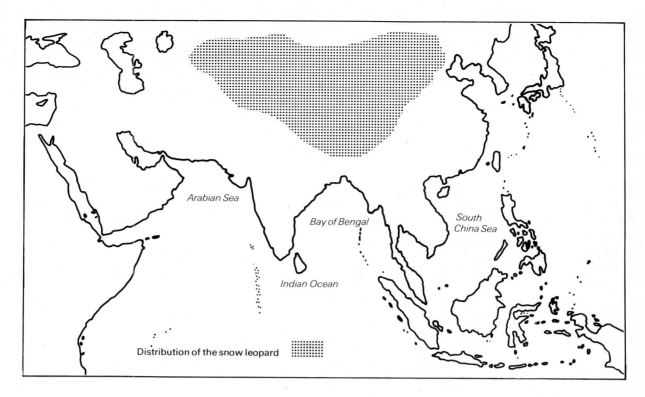

Distribution of the snow leopard

Until recently the demand for a snow leopard's pelt always outstripped the supply and it was profitable for local hunters to spend weeks, even months, stationed in the hills tracking a single animal. Various means of capture have been employed; for example, the Chinese bury a strong net under the snow, and by means of a trigger mechanism released by the animal's weight, the leopard is snared by becoming entangled in the net which is hoisted clear off the ground. This method prevents the valuable fur from being damaged. The animals are also trapped in deep pits covered over with vegetation and baited with a dead sheep. Perhaps the most ingenious means of capture is employed by the Bhotia tribesmen of Nepal who place spears tipped with a deadly poison concocted of local herbs along the leopard's well-worn trails.

In 1973 trade in the pelts of all spotted cats was banned by the member governments of the Washington Convention of International Trade in Endangered Species. This legislation has stopped the bulk of traffic of snow leopard pelts to Europe. Smuggling continues, however, and since 1973 the skins of these animals have somehow found their way in, especially to West Germany where the skins are made up into coats after which they can be legally exported to any country in the world.

The snow leopard is legally protected in India but this status has not helped protect it, and the same applies to Nepal where enforcement of the law is hampered by poor communications and limited manpower. In the USSR the snow leopard does find sanctuary in a couple of vast nature reserves in the southern province of Kazakhstan where there is an abundance of ibex and deer.

Estimates of present populations are difficult to calculate but from records of sightings and capture in all countries throughout its range there appear to be fewer than a thousand animals remaining in the wild. Approximately seventy are in various zoos including Helsinki and Chicago's Lincoln Park, both of which have had success at breeding this rare and beautiful animal.

RARE AND EXTINCT ANIMALS

Name	Status	Comments
European brown bear *Ursus arctos*	Extinct in Britain 10th century AD	Small numbers still hang on in many European countries, but has been hunted near extinction.
Aurochs *Bos primigenius*	Extinct 1627	Hunting and forest clearance caused the disappearance of this large wild ox.
Dodo *Rhaphus cucullatus*	Extinct 1681	Wiped out by sailors hunting it for meat and eggs, and by rats preying on eggs. One of 21 bird species destroyed in the Mascarene islands.
Guadeloupe parrot *Amazona violacea*	Extinct 18th century	Another island bird destroyed by man.
Wolf *Canis lupus*	Extinct in Britain 1743	Hunted for sport and as a thief of livestock.
Steller's sea cow *Hydrodamalis gigas*	Extinct 1768	Wiped out by sailors for meat and blubber only 27 years after its discovery.
Great Auk *Alca impennis*	Extinct 1844	A flightless penguin-like bird, killed by excessive egg collecting for food.
Quagga *Equus quagga*	Extinct 1878	Thought to compete with the introduced cattle of African settlers, and hunted to extinction.
Passenger pigeon *Ectopistes migratorius*	Extinct 1914	One of the most numerous of birds, with flocks of 2 billion, hunting and egg collection were responsible for its extinction.
Carolina parakeet *Conuropsis carolinensis*	Extinct 1914	The cypress swamps this bird lived in were cleared for orchards and the birds then shot as fruit thieves.
Heath hen *Tympanuchus cupido*	Extinct 1932	Hunting and habitat destruction decreased numbers to the point where the vigorous conservation programme introduced for the species was too late.
Caribbean monk seal *Monachus tropicalis*	Extinct 1952	Killed for fur and prevented from breeding by human disturbance, this species is almost certainly extinct.
Thylacine *Thylacinus cynocephalus*	Last zoo specimen died in 1930. A den was found in 1965	This animal known as the Tasmanian Wolf was shot as vermin by farmers, and unable to compete with introduced carnivores.
Java rhinoceros *Rhinoceros sondaicus*	1964 25 recorded 1975 45-54 recorded	The rarest of rhinoceroses, this species has made a slight recovery, but is still one of the most threatened of mammals and is sought by poachers for the supposed magical properties of its horn.
Mauritius kestrel *Falco punctatus*	10-20 individuals	Shot as a pest and further threatened by habitat destruction, this is the rarest surviving bird of prey. Conservation efforts may well be too late for it.

Name	Status	Comments
Californian condor *Gymnogyps californianus*	60 birds Declining	Hunted for sport and treated as a pest by farmers.
Arabian oryx *Oryx leucoryx*	Not seen in the wild since 1972 162 in breeding	Hunted for sport, sometimes by hunters using aeroplanes.
Walia ibex *Capra walie*	150 in 1967 300 in 1976	Destruction of the habitat is the main cause of decline It has also been hunted for meat and trophies.
Monkey-eating eagle *Pithecophaga jefferyi*	100 birds Declining	Forest destruction has caused decline, helped by live trapping.
Mediterranean monk seal *Monachus monachus*	500–1000 Declining	Killed as fisherman's competitor, also affected by over-fishing, pollution and disturbance.
Whooping crane *Grus americana*	40 in wild Small breeding group in captivity	The object of an intensive conservation programme which may be too late.
Aye-aye *Daubentonia madagascariensis*	Numbers not known, possibly only 30. 11 established in island reserve	One of many species in Madagascar threatened by habitat destruction.
Pere David's deer *Elaphurus davidianus*	Extinct in wild since 19th century 400 in zoo herds	Stable zoo population.
Indian lion *Panthera leo persica*	180 animals in reserve in NW India	This lion sub-species has been hunted and much of its forest habitat cleared.
Hawaiian goose *Branta sandvicensis*	Almost extinct in wild. Several hundred bred at Wildfowl Trust in England	Attempts to re-introduce captive-bred birds to Hawaii have had little success so far.
Orang Utan *Pongo pygmaeus*	5,000 animals Declining, but bred in many zoos	Young taken alive for zoos and as pets. Now strictly protected.
Kirtland's warbler *Denaroica kirtlandi*	1,000 birds In specially managed forest reserve	Can only breed in forests which are regularly burned. Fairly stable.
European bison *Bison bonasus*	Over 1,000 animals Increasing from low of 45 in 1920	Almost wiped out when the last large herd was destroyed in the first world war. Saved by captive breeding and reintroductions.
Sea Otter *Enhydra lutris*	Perhaps 6,000 in 2 areas. Recovering from low of 20-30 pairs in 1911	Hunted almost to extinction for its fur.
American buffalo *Bison bison*	10,000 In 1700 there were an estimated 60 million. By 1880 541 were left Now re-established	Slaughtered in millions for meat and in the establishment of farming.

Bibliography

THE GREAT AND THE SMALL

The Gorilla
Schaller, G. B., *The Mountain Gorilla, Ecology and Behaviour* (University of Chicago Press) Chicago 1963.
Schultz, A. H., *The Life of Primates* (Weidenfeld & Nicolson) London 1969.

The Etruscan Shrew
Brink, F. H. van Den, *A Field Guide to the Mammals of Britain and Europe* (Collins) London 1967.
Corbett, G. B. and Southern, H. N., *The Handbook of British Mammals* (Blackwell Scientific Publications) Oxford 1977.
Crowcroft, W. P., *The Life of the Shrew* (Max Reinhardt) London 1957.
Lawrence, M. J. and Brown, R. W., *Mammals of Britain, Their Tracks Trails and Signs* (Blandford) Poole, Dorset.
Southern, H. N., *The Handbook of British Mammals* (Blackwell Scientific Publications) Oxford 1964.

The Albatross
Jameson, W., *The Wandering Albatross* (W. Morrow) New York 1959.

The Hummingbird
Scheithauer, W., *Hummingbirds* (A. Barker) London 1967.
Skutch, A. F., *The Life of the Hummingbird* (Octopus Books) London 1974.

The Blue Whale
Ellis, R., *The Book of Whales* (A. Knopf) New York 1980.
Friends of the Earth, *Whale Manual '78* (Friends of the Earth Ltd.) London 1978.
Simon, N. and Géroudet, P., *Last Survivors* (Patrick Stephens) London 1970.
Slijper, E. J. *Whales and Dolphins* (University of Michigan Press) Ann Arbor 1976.

The Marmoset
Bridgewater, D. D. (ed.), *Conference on the Golden Lion Marmoset* (Wild Animal Propogation Trust) Washington DC 1972.
Hershkovitz, P., *Living New World Monkeys* (University of Chicago Press) Chicago 1977.
Walker. E. P., *Mammals of the World Vol 1* (The John Hopkins University Press) Baltimore, Maryland 1975.

The Giraffe
Dagg, A. I. and Foster, J. B., *The Giraffe, It's Biology, Behaviour and Ecology* (Van Nostrand Reinhold Company) New York 1976.
Guggisberg, C. A. W., *Giraffes* (A. Barker) London 1969.
Spinage, C. A., *The Book of the Giraffe* (Collins) London 1968.

FEROCITY AND VENOM

Poisonous Snakes
Bellairs, A., *The Life of Reptiles* (Weidenfeld & Nicolson) London 1969.
Minton, A. A. Jr. and Minton, M. R., *Venomous Reptiles* (Charles Scribner's Sons) New York 1969.
Morris, D. and Morris, R., *Men and Snakes* (Hutchinson) London 1965.
Schmidt, K. P., *Living Reptiles of the World* (Hamish Hamilton) London 1957.

The Polar Bear
Larsen, T., *The World of the Polar Bear* (Hamlyn) Feltham, 1978.

Pedersen, A., *Polar Animals* (George C. Harrap & Co Ltd.) London 1962.
Perry, R., *Polar Worlds* (David & Charles) Newton Abbot, Devon 1973.

The Killer Shark
Baldridge, H. D., *Shark Attack* (White Lion, Severn House) London 1977.
Budker, P., *The Life of Sharks* (Weidenfeld & Nicolson) London 1971.
Davies, D. H., *About Sharks and Shark Attack* (Routledge & Kegan Paul) London 1965.
Ellis, R., *The Book of Sharks* (Grosset & Dunlap) New York 1976.
Lineweaver, T. H. and Bakus, R. H., *The Natural History of Sharks* André Deutsch) London 1970.

The Box Jellyfish
Garnet, J. R. (ed.) *Venomous Australian Animals Dangerous to Man* (Commonwealth Serum Laboratory) Canberra 1968.

The Mosquito
Cloudsley-Thompson, J. L., *Insects and History* (Weidenfeld & Nicolson) London 1976.
Gillet, J. D., *Mosquitos* (Weidenfeld & Nicolson) London 1971.
Imms, A. D., *Insect Natural History* (Fontana) London 1973.

The Tiger
Corbett, J., *Man-Eaters of Kumaon* (Penguin Books) Harmondsworth, Middlesex 1979.
McDougal, C., *The Face of the Tiger* (André Deutsch) London 1977.
Mountfort, G., *Saving the Tiger* (Michael Joseph) London 1981.
Sankhala, K., *The Story of the Indian Tiger* (Collins) London 1978.
Seshadri, B., *The Twilight of India's Wild Life* (John Baker) London 1969.

The Vampire Bat
Linhart, S. B., *A Partial Biography of the Vampire Bat* (US Department of the Interior, Bureau of Sport, Fisheries and Wildlife) Washington DC 1971.
Turner, D. C., *The Vampire Bat* (John Hopkins University Press) Baltimore, Maryland 1975.
Yalden, D.W., *The Lives of Bats* (David & Charles) Newton Abbot, Devon 1975.

Poisonous Amphibians
Cochran, D. M., *Living Amphibians of the World* (Hamish Hamilton) London 1961.
Myers, C. W., Daly, J. W. and Malkin, B., 'A Dangerous Toxic New Frog Used by Embera Indians of Western Colombia With Discussion of Blowgun Fabrication and Dart Poisoning' *Bulletin of the American Museum of Natural History* Vol 161, Article 2, 1978.

The Nile Crocodile
Guggisberg, C. A. W., *Crocodiles* (David & Charles) Newton Abbot, Devon 1975.
Neill, W. T. *The Last of the Ruling Reptiles* (Columbia University Press) New York 1971.

The Scorpion
Cloudsley-Thompson, J. L., *Spiders and Scorpions* (Bodley Head) London 1973.
Savory, T. H., *Spiders, Men and Scorpions* (University of London Press) London 1961.

THE FAST AND THE SLOW

The Cheetah
Eaton, R. L., *The Cheetah: The Biology, Ecology and Behaviour of an Endangered Species* (Van Nostrand Reinhold) New York 1974.
Wrogemann, N., *Cheetah Under the Sun* (McGraw-Hill) New York 1975.

The Snail
Chinery, M., *The Natural History of the Garden* (Collins) London 1977.
Janus, H., *The Young Specialist Looks at Molluscs* (Burke Publishing) London 1965.
Kernly, M. P. and Cameron, R. A. D., *A Field Guide to the Land Snails of North-west Europe* (Collins) London 1979.
Taylor, J. W., *Monograph of the Land and Freshwater Molluscs of the British Isles* (Taylor Brothers) England 1894-1900.
Trueman, E. R., *The Locomotion of Soft-bodied Animals* (Edward Arnold) London 1975.

The Peregrine
Harris, J. T., *The Peregrine Falcon in Greenland* (University of Missouri Press) Columbia, Missouri 1979.
Ratcliffe, D. A., *The Peregrine Falcon* (A. D. Poyser) Berkhampstead, Hertfordshire 1980.

The Sloth
Goffart, M., *Function and Form in the Sloth* (Pergamon Press) Oxford 1971.
Dorst, J., *South America and Central America: A Natural History* (Hamish Hamilton) London 1967.

The Killer Whale
Cousteau, J., *The Art of Motion: The Ocean World of Jacques Cousteau* (Angus & Robertson) Brighton, Sussex 1973.
Lockley, R. M., *Whales, Dolphins, and Porpoises* (David & Charles) Newton Abbot, Devon 1979.
Mathews, L. H., *The Natural History of the Whale* (Weidenfeld & Nicolson) London 1978.
McDermott, J., 'Ocean Killers' *Wildlife Magazine*, June 1974.

THE SOCIABLE AND THE SOLITARY

The Cuckoo
Chance, E. P., *The Truth About the Cuckoo* (Country Life) London 1940.
Lack, D., *Ecological Adaptations for Breeding in Birds* (Methuen & Co Ltd) London 1968.

The Seal
Hewer, H. R., *British Seals* (Collins) London 1974.
King, J. E., *Seals of the World* (British Museum of Natural History) London 1964.
Maxwell, G., *Seals of the World* (Constable) London 1967.
Scheffer, V. B., *Seals, Sea Lions and Walruses* (Stanford University Press) Stanford 1958.

The Mole
Godrey, G. and Crowcroft, P., *The Life of the Mole* (Museum Press) London 1960.
Mellanby, K., *The Mole* (Collins) London 1971.

The Penguin
Richdale, L. E., *Sexual Behaviour in Penguins* (Kansas University Press) Kansas 1951.
Simpson, G. C., *Penguins* (Yale University Press) New Haven, Connecticut 1976.
Stonehouse, B., *The Biology of Penguins* (Macmillan) London 1975.

The Baboon
Altmann, S. A. (ed.) *Social Communication Among Primates* (University of Chicago Press) Chicago 1976.
Morris, D., *Primate Ethology* (Weidenfeld & Nicolson) London 1965.
Rheingold, H. L. (ed.) *Maternal Behaviour in Mammals* (J. Wiley & Sons) New York, 1963.

Schultz, A. H., *The Life of Primates* (Weidenfeld & Nicolson) London 1969.

The Solitary Wasp
Evans, H. E., *The Comparative Ethology and Evolution of the Sand Wasps* (Harvard University Press) Harvard 1966.
Evans, H. E. and Eberhard, M. J. W., *The Wasps* (David & Charles) Newton Abbot, Devon 1973.
Fabre, J. H., *The Hunting Wasps* (Hodder & Stoughton) London 1919.
Spradbery, J. P., *Wasps* (Sidgwick & Jackson) London 1973.

THE GAUDY AND THE CAMOUFLAGED

The Frigate Bird
Austin, O. L. Jr., *Birds of the World* (Hamlyn) Feltham, Middlesex 1970.
Burton, J. A., *Birds of the Tropics* (Orbis Publishing) London 1973.
Harrison, C. J. O. (ed.) *Bird Families of the World* (Elsevier-Phaidon) Oxford 1978.
Landsborough-Thompson, Sir A. (ed.) *A New Dictionary of Birds* (Nelson) London 1965.

The Chameleon
Grzimek, B., *Animal Life Encyclopedia Vol 6* (Van Nostrand Reinhold) New York 1975.

Insects Resembling Plants
Cott, H. B., *Adaptive Colouration in Animals* (Methuen & Co Ltd) London 1940.
Fogden, M. and Fogden, P., *Animals and Their Colours* (Peter Lowe) London 1974.
Ward, P., *Colour for Survival* (Orbis Publishing) London 1979.

Snake Mimics
Wickler, W., *Mimicry in Plants and Animals* (Weidenfeld & Nicolson) London 1968.

The Octopus
Lane, F. W., *Kingdom of the Octopus* (Jarrolds) Norwich, Norfolk 1957.
Morton, J. E., *Molluscs* (Hutchinson) London, 1971.
Wells, M. J., *Brain and Behaviour in Cephalopods* (Heinemann) London 1962.
Wells, M. J., *Octopus: The Physiology and Behaviour of an Advanced Invertebrate* (Chapman & Hall) London 1978.

Coral Reef Fishes
Bennett, I., *The Great Barrier Reef* (Lansdowne Press) Melbourne 1971.
Carcasson, R. H., *A Field Guide to the Coral Reef Fishes of the Indian and West Pacific Oceans* (Collins) London 1977.
Fricke, H. W., *The Coral Seas* (Thames & Hudson) London 1973.
Ormond, Dr. R., 'Deceptions on the Coral Reef' *New Scientist* March 1981.

Nightjars, frogmouths and potoos
Grzimek, B., *Animal Life Encyclopedia Vol 8* (Van Nostrand Reinhold) New York 1972.

Birds of Paradise
Gilliard, E. T., *Birds of Paradise and Bower Birds* (Weidenfeld & Nicolson) London 1969.
Rutgers, A., *Birds of New Guinea* (Methuen & Co) London 1970.

A SHORT LIFE AND A LONG ONE

The Giant Tortoise
Flower, Maj. S. S., Further Notes on the Duration of Life in Animals *Proceedings of the Zoological Society of London* Series A, 1937.
Minton, S. A. Jr. and Minton, M. R., *Giant Reptiles* (Charles Scribner's Sons) New York 1973.

The Housefly
Oldroyd, H., *The Natural History of Flies* (Weidenfeld & Nicolson) London 1964.
Sanders, E., *An Insect Pocket Book* (Oxford University Press) London 1946.

The Robin
Lack, D. L., *The Life of the Robin* (H. F. & G. Witherby) London 1965.
Sharrock, J. T. R., *An Atlas of Breeding Birds in Britain and Ireland* (A. D. Poyser) Berkhamstead, Hertfordshire 1976.
Witherby, H. F. (ed.) *The Handbook of British Birds* (H. F. & G. Witherby) London 1965.

The Axolotl
Bates, M., *The Nature of Natural History* (Charles Scribner's Sons) New York 1950.
Garstang, W., *Larval Forms with Other Zoological Verses* (Blackwell Scientific Publications) Oxford 1962.
Ley, W., *Salamanders and Other Wonders* (Viking Press) New York 1955.
Smith, H. M. and Smith, R. B., *Synopsis of the Herpetofauna of Mexico Vol 1, 2* (E. Lundberg) Augusta, West Virginia 1971-73.

The Queen Termite
Harris, W. V., *Termites: Their Recognition and Control* (Longman) London 1971.
Hicklin, N. E., *Termites: A World Problem* (Hutchinson) London 1971.
Howse, P. E., *Termites: A Study in Social Behaviour* (Hutchinson) London 1970.

The Asian Elephant
Blond, G., *The Elephants* (André Deutsch) London 1962.
Carrington, R., *Elephants* (Basic Books) New York 1959.
Sillar, F. C. and Meyler, R. M., *Elephants Ancient and Modern* (Viking Press) New York 1968.

The Dragonfly
Chelmick, D., *The Conservation of Dragonflies* (Nature Conservancy Council) London 1980.
Corbett, P. S., *Dragonflies* (Collins) London 1960.
Hammond, C. O., *The Dragonflies of Great Britain and Ireland* (Curwen Books) London 1977.
Longfield, C., *Dragonflies of the British Isles* (F. Warne) London 1949.

THE NUMEROUS AND THE RARE

Mice and Rats
Hanney, P. W., *Rodents: Their Lives and Habits* (David & Charles) Newton Abbot, Devon 1975.

The Monkey-eating Eagle
Bonitt, C. B. Rundquist, L and Rundquist, V. M., *The Monkey-eating Eagle of the Philippines* (Department of Natural Resources, Bureau of Forestry Development) Davao City, Philippines 1975.
Brown, L. and Amadon, D., *Eagles, Hawks and Falcons of the World* (Country Life Books) Feltham, Middlesex 1968.

The Desert Locust
Centre for Overseas Pest Research. *The Desert Locust Pocket Book* (Centre for Overseas Pest Research) London 1978.
Uvarov, Sir B., *Grasshoppers and Locusts* (Cambridge University Press) Cambridge 1977.

The Arabian Oryx
Jungius, H., 'Plan to Restore Arabian Oryx in Oman' *Oryx: Journal of the Fauna Preservation Society* London 1978.

The Quelea
Major, J. and Ward, P., *Illustrated Description, Distribution Maps and Bibliography of the Species of Quelea* (Centre for Overseas Pest Research) London 1972.
Ward, P., *The Role of the Crop Among Red-billed Quelea* (Ibis 120) London 1978.

The Snow Leopard
Jackson, R., 'Snow Leopards in Nepal' *Oryx: Journal of the Fauna Preservation Society* London 1979.
Roberts, J. T., *The Mammals of Pakistan* (E. Benn) London 1977.

The Tasmanian Wolf
Ovington, D., *Australian Endangered Species* (Cassell) Stanmore, New South Wales 1978.
Serventy, V., *A Continent in Danger* (André Deutsch) London 1966.

Author's Acknowledgments
The author would like to thank Michael Allaby, David Sharp and Elizabeth Dron for help and contributions to this book.

Index

Figures in italic refer to illustrations

207